Praise for
Birds, Beasts and Bedlam

'Derek's passion for our native wildlife is matched by his highly pragmatic approach to helping it thrive. He gets stuff done. The story of his amazing work at Broadwoodwidger is highly entertaining but also a heartfelt plea for a wilder, more inspiring Britain.'

HUGH FEARNLEY-WHITTINGSTALL

'Derek Gow's riotous adventures rescuing threatened species and releasing them for rewilding read like Gerald Durrell on steroids. Courageous, visionary, funny and always up for a scrap with bureaucracy and complacency, the world needs many more Derek Gows.'

ISABELLA TREE, author of *Wilding*

'A brilliant read – and the entertaining backstory of a species of human megafauna who has transformed the British conservation scene.'

BENEDICT MACDONALD, author of
Rebirding (winner, 2020 Wainwright Conservation Prize)

'A great read from a rewilding polymath. This is how it's done – and you'll also learn about the struggles. Nature needs more bold people like Derek Gow to restore our damaged planet.'

ROY DENNIS MBE, author of *Restoring the Wild*

'A larger-than-life character who writes as well as he thinks, Derek Gow has seen the future – rewilding – and it works.'

STANLEY JOHNSON

'There is only one Derek Gow. Like a gruff, bearded naiad, he speaks for nature with all the force of someone who has spent a life protecting it. The trickle of good news stories coming out of British conservation nearly all owe a significant debt to Gow; he has been instrumental in the restoration of the marvellous beaver, the increase in water voles and the return of storks to English land. In *Birds, Beasts and Bedlam*, this fascinating man tells us about his life and how he ended up championing rewilding as a solution to our impoverished landscape. More than this, there are hilarious stories of his many interactions with animals, both wild and less than pleased, that have dotted his journey in turning sterile Devon farmland into a beautifully rewilded tapestry of faunal interactions.'

DR ROSS BARNETT, author of *The Missing Lynx*

'It is a charming, passionate and timely book. It will stir thoughts in many, and motivate them to do even small things that can have large consequences. I hope it will become a classic.'

BERND HEINRICH, author of *A Naturalist at Large*

'Derek Gow's *Birds, Beasts and Bedlam* is charming, witty and has a "get it done" approach to the reintroduction of endangered species and restoration of natural habitats destroyed by hundreds of years of overmanagement.'

BENJAMIN KILHAM, wildlife biologist;
author of *In the Company of Bears*

Birds, Beasts and Bedlam

*Turning My Farm into an Ark
for Lost Species*

Derek Gow

Chelsea Green Publishing
White River Junction, Vermont
London, UK

Project Manager: Patricia Stone
Developmental Editor: Muna Reyal
Copy Editor: Susan Pegg
Proofreader: Angela Boyle
Indexer: Linda Hallinger
Designer: Melissa Jacobson
Page Layout: Abrah Griggs

Printed in the United States of America.
First printing April 2022.
10 9 8 7 6 5 4 3 2 1   22 23 24 25 26

Library of Congress Cataloging-in-Publication Data
Names: Gow, Derek, author.
Title: Birds, beasts and bedlam : turning my farm into an ark for lost species / Derek Gow.
Description: White River Junction, Vermont : Chelsea Green Publishing, [2022] | Includes
    bibliographical references and index.
Identifiers: LCCN 2022003443 (print) | LCCN 2022003444 (ebook) | ISBN 9781645021339
    (hardcover) | ISBN 9781645021346 (ebook)
Subjects: LCSH: Gow, Derek. | Wildlife reintroduction—Great Britain. | Wildlife
    conservationists—Great Britain—Biography.
Classification: LCC QL83.4 .G69 2022 (print) | LCC QL83.4 (ebook) | DDC 639.970942--dc23/
    eng/20220131
LC record available at https://lccn.loc.gov/2022003443
LC ebook record available at https://lccn.loc.gov/2022003444

Chelsea Green Publishing
85 North Main Street, Suite 120
White River Junction, Vermont USA

Somerset House
London, UK

www.chelseagreen.com

To Kyle and Maysie; the best kids ever.

— CONTENTS —

The Great False Idol of the Industrial Machine

WE COULD SMELL THEM FROM BEHIND THE large fallen tree to which we had belly crawled. Their warm sweet breath. Their musky bovine richness. Slowly we extended our heads up to get a view. They were vast. Simply gigantic. Swimming in a sea of lush understory, they wrenched and pulled at the vegetation around them, snapping twigs and grinding their woody repast like trying teenagers crunching dry breakfast cereal. Humps extended, they reached up into the foliage to pull down dainties with their dark curling tongues. Bark and berries, leaves and shoots. They snorted, stamped and shook their rough maned heads with bleary annoyance when buzzing bloodsuckers bit. Swishing tails and dun coats with deep winter wool already forming, the Polish bison bulls were breathtaking.

We watched for a while. When we were done, my guide said it was best that they knew where we were. We stood up slowly and clapped our hands. Their heads shot up in an instant, snorting to catch our scent, chestnut eyes wild with their whites showing clear. Nostrils extended. They whirled and plunged, tails up in alarm, into the backing infinity of deep forest green. Branches swished, sprung and steadied. In the haste of flight, the splatter from their steaming dung dripped slowly from branched limbs and leaves to the ground.

Although there is no evidence that the European bison (*Bison bonasus*), known as wisent (pronounced 'we-sent'), ever occurred on our islands, as a hybrid of the extinct steppe bison (*Bison priscus*) and the aurochs (*Bos primigenius*), both of which did, they are attuned to our environments. Britain once hosted a broad range of great beasts. We slaughtered the bears, elk and lynx many centuries ago. The wolves lasted longest. Now, only the names of their crags, hills, meres or the ubiquitous deep pits where we caught and bound them for torture recall their once being. Like the aquamarine blue moor frogs, black storks and night herons, we were the end of them all.

One in seven of our surviving species is now also threatened with extinction. In large part, much of the landscape so seemingly green, which we traverse with daily indifference, is dead. Chemicals and pesticides in the soil have killed the very small things. The passing of these has so starved, poisoned or otherwise compromised the other slightly larger, but still small things, that shockwaves of depletion now ripple upwards through every level of natural dependency. Gone is the food for some creatures or the cover for others. The living space that remains is highly restricted and commonly of poor quality. The absence of one pivotal creature can mean the loss of natural function upon which others depend. Even when our understanding of this is crystal clear, we act in reluctant slow-motion response.

———

Conservation comes in many forms and my beginning was not with the wild but the tame. At a time when you can drive through the landscape and see so many of the old black or spotted sheep, white long-horned cattle, or brick-red pigs more or less everywhere, it's hard to remember that these relict sorts were by the 1970s nearly extinct. Farming at that time was already set to conquer its Everests of 'improvement'. Rivers of government cash flowed into subsidises for everything imaginable. The import of faster-growing continental livestock, new and super productive crops, fertilisers that flowed from white plastic sacks rather than freely from cows' backsides, pesticides that killed their target species and much more besides. Guilds of focused advisors in drab brown overalls and tiny vans met farmers free of charge to explain how to employ this largesse. Colleges produced legions of indoctrinated students who marched out in ranks to feed the world. Research stations, laboratories and experimental farms, all centrally funded, were established throughout the land. Meadows full of dancing wildflowers or woodlands where spotted fly-catchers dipped and weaved to catch beakfuls of insects twirling in sunlit strobes did not fit the narrative of those times. Most were ploughed under or ripped free from the soil that had held them for centuries for incineration on pyres well prepared. Birds of all sorts died in myriads when cornfields, old pastures and orchards were sprayed with new toxins. Frogs returned to breed in the spring to ancestral ponds now filled in. There are pictures of them in black and white, sitting in massed aggregations on their drying spawn with no water.

The photographers who took these images wept.

Breeds of livestock with their roots buried deep in Britain's culture were discarded as well. It did not matter that they had adapted to frugal living to produce something – a little meat, milk, horn or dung to fertilise small fields – for folk who had nothing and could offer them less. Who cared if they had been brought by the Norse or the Romans or the Celts? They were out of time. Small or slow growing. Difficult to handle with independent spirits. The sooner they were all gone, the better. Their other qualities of disease resistance, fine wool or super-lative meat meant nothing. Any adaptation to specific environments was meaningless in a time when whole landscapes could be rearranged.

To be clear, as individuals I like farmers very much. It's the great false idol of the industrial machine that so many unblinkingly worshipped that's the problem. In the main they are a well-humoured bunch. The old ones with the good stories always are best, and I have spent many hours sitting in their cosy kitchens listening to their tales where small dogs snoozed next to Agas and busy wives bustled to serve cakes. There was slight Henry Cowan who regretted till the day he died that he'd allowed a passing dealer to buy his last two horses, kept long after the others had gone, for the glue works. Tall Francis Watson, a big bear of a man who at the age of seventeen had guarded the palace of the Nizams in Hyderabad and whose great joy it was to linger for no particular purchase in our village shop to converse with its Pakistani proprietors in Urdu. Slight Miss Bartholomew whose old cats pissed on her house chairs and whose ancient pet pigs were turned by her stockman daily to ease their bed sores when they could no longer stand. All once of great colour who have passed now in time.

Their world was simpler, of clear rights and dark wrongs. The reapers who harvested in their golden youths are not of the sort that scythe the earth today. The prospect that the land that they had cleared of rocks and drained and deforested, and then reforested and enriched and impoverished in the swiftest succession, would ever be used again for any purpose other than farming would not to them have seemed plausible at all. The notion that some of the oldest beasts could be restored to accelerate nature's gain would have seemed absurd.

So why bother to bring back bison to Britain when we could be content to sit back in our slippers to reintegrate beavers into the countryside, which, in theory at least, is as easy as falling off a stationary bus? The answer in large part is process. If, as it seems tantalisingly tangible, we are going to move from an era of unequivocal public subsidy for farming 70 per cent of the British landmass (23 million acres) in some form or another to a time when public money will be employed more evenly to repair nature, then at least a few of the large creatures we hunted to extinction may be restored in a limited fashion to assist this endeavour. Bison, for example, are not cattle. They are high forest browsers. If you reinstall them in dark, dull plantation woodlands with little biodiversity value, they will smash and debark big trees, wallow in sand soils, gouge out damp clays, provide pesticide-free blood and dung in abundance for insects, and crunch down woody scrub at random in a jagged and irregular manner.

The bark they rip from the stems of broad-leafed trees in a frozen winter by inserting the teeth of their lower palate under its surface, gripping it tight with their upper jaw and tugging sharply downwards will 'whip crack' the length of the stem before it tumbles downwards like a falling curtain to be consumed. A single bison can eat thirty-two kilos of bark in a day. Multiply this by a stamping herd, hoar frosted with steaming nostrils, and their impact on woodland structure becomes obvious. Whole groves of succulent, young trees are retarded or misshaped. Their wounds leach resin or sap, which snails cluster in to exploit. Some bare areas may scab and scar over while others decay completely for woodpeckers to pock full of voids. Bats, martens and birds use these cavities as nesting sites while specialists such as willow tits make their own abodes in desiccated pockets rotted down by mycelia of many sorts. Nature loves random and there is more in the simplest of forms. The fur from a bison's woolly coat will be gathered by birds from the grasping thorns of bramble or rose, or from their backs directly when it peels in scrofulous mats in the spring time. This warm snuggly material, which is ideal for their nests, will be filched from them in turn by small mammals to take underground. The repetitive wallowing of bison in dry sand banks scours these features free of vegetation in random patches. In their well-trampled base lie easily excavatable egg-laying areas for sand lizards, while mining insects pit with their tunnels any exposed standing banks. Over time, there is always the fragrant possibility that the child-painted wonder of yellows, blues, browns and greens that is the European bee-eater will one day grace them as sites for their nest tunnels.

Bison will, in short, do some things that cattle are not capable of doing and others that cattle don't do very well. This of courses is hardly surprising, given that ten thousand years' worth of preparation for domestication has profoundly altered the shape, biology and behaviour of cattle, while bison have retained their wild being intact.

For all these reasons, herds of wisent already roam Dutch nature reserves, such as the sand dunes of Zuid-Kennemerland National Park within easy reach of

Amsterdam. The visionary thinking behind projects of this sort in Europe is well over half a century old. With flare and imagination, they are not hard to accomplish and could easily become a British reality.

Large wild herbivores running around developed human landscapes are, however, a problem. If you think not, then the figure of around twenty human fatalities from collisions with approximately seventy-four thousand deer on UK roads annually may persuade you otherwise. If bison are to free live again in Britain, it will either be in the kind of large 205 hectare forest pens proposed for West Blean woods nature reserve near Canterbury, in a partnership between the Kent Wildlife and Wildwood Trusts, or perhaps at a time when technology controls their ranging with electronic neck collars, grazing tall around the standing sarsens of Stonehenge.

———

In my time, I have tried hard to save some things from slipping away. Rotund furry water voles with their black, beady eyes, cinnamon-yellow dormice, ball-nesting harvest mice with curling prehensile tails and rust-red squirrels. All of these are now lost from much of their former range and are continuing to fade fast. Over time, I have kept and bred in captivity most of the mammal species that belong on our islands and a random spectrum of others that got here and otherwise do not. It's been an absorbing experience and, although some of the knowledge I have acquired is of dubious relevance (did you know that tame roebucks can't gore you if you cut off their bone-hardened antlers once their rich velvet has shed and fix short sections of garden hose to their stumps with jubilee clips or that hand-reared brown hares can jump as high as waist height to bite you firmly in the groin?) I have nevertheless learnt a lot. I have bred many creatures which were once considered common and realise full well that their existence in abundance is no longer the case. Long years after his death, the ebullient zookeeper Gerald Durrell's vision of using captive breeding to save endangered species has become a mantra for zoos worldwide. That the route he espoused is not easy or straightforward does not mean it's wrong, it's just not as simple as he understood in his time. While for some creatures the circumstances that diminish their being are easy to fix – just stop killing them as a whole society and they will bounce right back when you put them in suitable environments or make new space using natural architects such as cattle and bison, beaver and ponies or boar and water buffalo – others are not so simple. The grey-breasted corncrakes with their short pink bills and barred brown backs, whose repetitive rasping calls were once so ubiquitous that they stopped country dwellers from sleeping on warm summer nights, are silent because the vast hay meadows grown to feed the working horses are gone. The insects that filled the hay crop full have gone. The untidy countryside that left random rough edges in plenty has gone. Untidy corners are few and, even where these exist on a scale that seems large like the Nene Washes in Cambridgeshire, every predator that can consume them is hunting there for food. The corn-crakes' own short lives are complicated by long annual migrations, which ensure that they are exposed to a multiplicity of further hazards. Although projects to sustain them, such as that developed by the Royal Society for the Protection of Birds in the western isles of Scotland, show that you can collaborate with sympathetic farmers, they will require detailed cultivation if they are to survive.

Thirty years on, I am working with some of Durrell's disciples in an effort to restore species such as red-backed shrikes, white storks, glow-worms, beavers and wildcats. Together with the fine folk who work with me, I have helped to advance a case for the restoration of others like the eelpout or brown marbled burbot – a torpedo of a fish with a single chin whisker – or the dapper dalmatian pelican with its bright yellow beak and bouffant head curls. Other individuals of great worth have fought their own battles to ensure that coal-black choughs with curving red beaks birl in the skies over Jersey or to enable a growing flock of barrel-bodied bustards to strut in a military manner across the grasslands of Salisbury Plain. The champions of curlews and cranes, of kites and godwits, or for the last of the sad pearl mussels confined as a population to a single Welsh fridge, are all truly remarkable people.

I have reformed my farm of 300 acres on the wet, windy Cornish border with West Devon into an independent wildlife centre where wild creatures of any sort required can be bred in captivity for further study and released back into areas of suitable habitat in time. The land itself is being rewilded to enable any life that has survived to recover from farming, if it can. While many other individuals and organisations are attempting the same, it is nevertheless sobering that, in this time of near miracles, when reintroduced white-tailed eagles and ospreys soar with growing confidence above southern seas, every graph there otherwise is of natural life resembles the trajectory of a *Thunderbirds'* rocket that has run out of fuel.

Plummeting in an accelerating arc of downwards destruction.

It does not have to be this way. We know and can accomplish so much.

This is the story, in large part, of my own life journey (which is, I earnestly hope, not quite over yet), from breeding endangered breeds of domestic farm livestock at its beginning to restoring a broad array of the most marvellous native creatures back into habitats they have lost at its end.

CHAPTER ONE

It Was a Wildling that Wanted Away

WHEN I WAS SMALL, I WAS GIVEN A SHETLAND ewe for my birthday by a farmer friend who wished, as they do, to foster in male children from agricultural communities a fanatical interest in sheep. She had a dark brown body, a white face with black patches over her eyes and, for her breed, the unusual feature of tight curly horns on the top of her head. She was in lamb to a purebred ram and, as she was multicoloured, I was informed could well produce offspring of unpredictable patterns in the spring.

That possibility enthralled me.

The Shetland isles, despite being British, have never abandoned their long love affair with Scandinavia. The coloured sheep that graze there are named after the landscapes and textures that the Norse shepherds saw and, to this day, tones such as moorit (moor red) – a sort of warm ginger tan – or shaela (hoar frost grey) – a black sheep with grey tips to its wool – are tinged with the vision of far northern eyes.

Perhaps my ewe longed for her windswept pastures because she never tamed. Despite care and tempting foodstuffs, whatever field she was put in with whatever combination of other sheep, she always escaped. In a final bid to restrain her wandering, my mum tethered her one morning to a metal stake on a well-grassed slope next to our house. She disliked this arrangement immensely and spent the rest of the afternoon racing from one end of her chain to the next. Assured by others that she would settle to a more sedate pace of existence once she realised her limits, I went to bed. In the morning she was dead, with the chain curled round her neck and her tongue extended in a last great blueing gasp of suffocation. Hard, rigid with her legs contorted and head thrown back, she was a wildling that wanted away.

There is an old farming saying that where there is 'livestock there is deadstock'. It's true and the many years of living with or caring for very many more animals of all sorts since has inured me time and again to its reality. I can't remember the warm salt tears that flowed when I found this first one gone, though they must have freely fallen. The pain of losing her with its death of illusory hopes and ambitions hurt so very much.

I lived in a white house on the edge of a golf course in the tiny market town of Biggar in the Scottish borders with my mother, brother and gran. Our home overlooked the Biggar Burn in the distance as it wove its long course to the River Tweed, along the foot of the Culter Hills. Its wide, wet margins, which attracted great skeins of pink-footed geese from Iceland in the autumn, were filled with peewits and curlews calling in the spring. Birdsong was so common in those days that it was indivisible from air. Green those hills were. The dark grim lines of the conifer plantations, which reached from their spine away down to Moffatt on their route to the border, did not sully our view.

Biggar was a market town and, like many others at that time, would awake on a Saturday morning to the sounds of cattle lorries, trailers and trucks. Their bleating and bellowing cargos, shouting farmers and barking dogs in crescendo rose week on week to reach fever pitch in the autumn when the hills disgorged their annual avalanche of sheep and cattle. Lambs that were not required to maintain the breeding stock of the high hirsels where they were born, ewes too old to keep as their teeth slackened and fell free, surplus rams with great whorled horns, whose growth rings, ridged like an ebb tide's sand, told their age.

Mostly they were the black-faced breed, the old Linton sheep kept in the borderland from a time long forgotten that were encouraged to come north by the agents for the highland chiefs to afford a cash income from their vast but poor estates. This breed, as well as the 'long sheep' or white-faced Cheviot from a softer hill range to the east, displaced not only the small aboriginal flocks of multi-horned dark sheep that the chiefs' tenants kept but also, in the end, the people themselves. Wool and mutton were more valuable by far than any rent that crofters could provide and, after the savage defeat of Culloden in 1746 smashed the old system of clan inter-dependency between individuals, southern sheep with southern shepherds for the profit of the farmers who owned them and the chiefs who leased the grass became the new world order.

Lowland breeds kept in the flatlands of the valley bottoms, bluefaced or border Leicesters, Suffolks and their hybrid 'mules' were also brought into the Biggar market. Continental imports such as the Dutch Texels, French Charolais or pink-nosed Finnish landrace sheep in the 1970s were rare. This last breed created considerable excitement at one time because they can produce up to five lambs at a single birth. Although government-sponsored animal breeding centres with salaried staff who went home for dinner on time and thus knew little of farming championed their utility, every shepherd with sense soon realised that a single live lamb from a blackface was worth way more than a whole litter of dead ones from a Finn. They were simply too soft to survive on the cold, damp pastures of our hills and their tendency to produce offspring well before the spring grass came was lethal.

I was perhaps twelve when I began my part-time job in the market. A shrill, sharp, smelly kid who ran with others of the same sort in a pack with the limping market dogs up and down gated alleys chasing sheep. Our working day started early in the morning when the sun rose sufficiently to go out to the fields to gather the bunches delivered in days before. These flocks, which were quite typically mixed dealers' batches – a hodgepodge of breeds and sexes bought together cheaply from other auctions earlier in the week to be pedalled at ours for a hopeful quick profit – were run along footpaths at stream sides or lanes that had been used for this purpose for centuries. Some of us would race in front to block alleys and other escape routes. Most properties we passed had good wrought-iron fences with sound latching gates to protect their gardens. Those who did not took their own chance when the sheep scattered and ran like a freight train through verdant borders of delicate flowers and ordered root vegetables. No one in those days considered it abnormal to move free livestock through a village or town, or along a main street. It had been that way forever. On arrival in the market the flocks would be run into a shedder. This old wooden system of black tarred gates and alleys, with many separate pens leading off a central narrow race and with free-opening gates in its top or middle sections, was designed to ensure that a couple of people could separate a large complicated assortment into pens of a uniform order swiftly. While some of us whooped and waved plastic feed sacks to chase the sheep up the race, where they were quickly separated by the gate operators, others pulled out odd individuals.

A few rams would be penned, singly, nearer the ring. Some old, poor cull ewes would go straight to the abattoir to avoid the gaze of any animal-welfare inspectors.

Order was all in the presentation. No black-faced sheep with white-faced ones. All grey-faces together. Black or spotted sheep on their own.

Thus sorted, they were chased into numbered pens for the buyers to peruse. Some jumped in to check their teeth and udders to ensure that any breeding sheep were sound and without complicated lumps that meant their milking days were done. Others felt the back fat where it covered their middle spine. Plump sheep were destined for a swift end in the days that followed. Bought on a Saturday in the borders of Scotland, they would be dead on a hook in a Birmingham chiller by first thing on Monday. Young sheep with their full life ahead, good breeding rams and middle-aged ewes would be presented for sale on a well-polished basis. Fresh smelling of coal tar from the coloured dip used to turn their wool ochre or a broad range of yellows, they might sport a neat horn brand, coloured ring of electrician's tape or tiny red numerals neatly placed with oiled paint on the cheek skin of their face. It was believed quite utterly that this ornamentation 'stood them out' to catch the buyers' eyes. The large rams, trimmed to tidy perfection, would have numbers painted on their tabletop backs. When the sale began, sheep would be run up to the main ring. Farmers of all ages and attires, in clothes new and old, would lounge and lean. Bids would be taken by the auctioneer on the basis of winks, finger movements or slight nodding heads. When no further advance was offered, the lot would be sold and off, they would trot to their fate of the day.

It Was a Wildling that Wanted Away

My own sheep flock started on a low-budget basis. Sometimes in the arriving groups would be ill-grown lambs whose mothers had perhaps died or failed to milk well. For whatever reason, these tiny effigies of their well-grown kin were of little worth. Sometimes a shepherd would give me one free in return for helping or I would purchase another for a risible sum. I kept these in an old pony paddock, which a pal of my grandmother's owned, in an effort to grow them to breeding age. It was a fight against reality. Virtually all were stunted and poor. Most died within a short time and those that did live made little profit. I persevered for years to no point.

Then, one day in the late 1970s, more Shetlands appeared in the market. Exhausted from a long two-day journey down from the isles and turned green by their own liquid shit, they were small, rough and ragged. They did not cost much, but when I took them home and put them out in my lush grass paddock, they did well. In the spring, some had lambs with gold and ginger patterns. A lifelong love affair with old breeds of domestic livestock was born with their offspring. I loved them as they grew and put on weight. As their wool lengthened and their tiny horns emerged. Their confident battles on mounds with their friends. Their jumping and games and glorious sunset races.

Incrementally their numbers grew, and I rented more poor land for grazing: old railway tracks and ex–horse paddocks, small pens behind a local abattoir. A multi-horned Manx ram came from Dumfries, more moorit

Shetlands were delivered from Haltwhistle one day when I was absent and were left with their legs bound by their deliverer in my grandmother's coal shed. Jet-black Hebrideans, black-and-white spotted Jacobs, dun-faced Lintons from the main Isle of Lewis. Some Borerays abandoned by the St Kildan islanders when they fled their island archipelago in the 1930s and a few small deer-like Soays, whose name means 'sheep isle' in Old Norse, with white belly and tail patches who were likewise marooned. Feral Ronaldseys came from the seaweed shores of the distant Orkneys and promptly escaped. Used to being isolated on wrack stands by the tide, they could, to my surprise, swim well and considered water as no barrier at all. When I ran the last to ground in a stream that flowed by a public footpath and a righteous woman in tweed asked what I was doing, my response that it was an 'aquatic sort of sheep that I was teaching to swim' was reported back to displease grandmother in an instant.

My fences were not good and, when I was working, mother would commonly receive notice that one or other of my oddities had made another bid for freedom. Though some of these happenings were funny, most were not. Excuses as to why 'Mini-lamb' – another bottle-reared Orcadian – had once again consumed a garden full of roses or been captured looking into the baker's window in the high street amused her not at all. The worst was when a Shetland ram with whorled horns jumped into a neighbouring farmer's flock of pedigree border Leicesters. Although gargantuan, the farmer's rabbit-faced lothario of a male was no match for this dapper, swift and unexpected rival. To state that the resultant crop of medium-sized ginger lambs that his pedigree ewes produced the following year caused their owner consternation would be an understatement, and he came looking for me. That meeting remains one of the most uncomfortable experiences of my life. He swore, he shouted and, when I offered to pay, he laughed. He knew I could not afford to do so. It was pitifully humbling and awkward. Mum transported the Shetland ram swiftly to the slaughterhouse as a sacrificial offering to his rage.

The summer months were filled with agricultural shows. Together with a small band of like-minded enthusiasts, I would preen my finest sheep to ensure the best impression. Trimming would take many evenings of hours to achieve an even coat finish with no straggling ends. Using hand shears, you had to take just enough off to ensure smooth uniformity, but not so much that dips and hollows appeared in a coat that you wished to be even. Black breeds would have their dark horns, legs and faces oiled till they glistened. Light ones would be washed with soap powder and then blown dry. Competitors would assemble on a given date in the early morning at a temporarily tented settlement surrounded by serried rows of wooden or metal pens. Peebles, Dalkeith, Biggar and Abington shows were all chummy local events. Everyone competing knew everyone else, and few really cared about win or lose. While rosettes were nice, a happy afternoon in the beer tent with chums was much more fun. At a time when drink-driving was not the unacceptable behaviour it is now, great indulgence in alcohol was a principal mainstay of the gathering. One show ran a competition year on year for the drunkest farmer on the showground. A family called Dunn won it on nearly every occasion, and when their next victory was announced, would duly stagger, be carried or driven, laid out in the bed of a long wheelbase Land Rover, to the main ring to collect their prize. It came in two parts. The first was a near human-sized bottle of whisky. The second was that the local police, after another round of congratulatory drinks had been absorbed, would personally put the winner in their car and follow him weaving his way homewards to ensure no accidents occurred.

Wildlife remained all around. Elfin red squirrels flitted along the back dyke of my father's garden in Broughton from the larch plantations to the west. They stole plums in season and some of the neighbours shot them. Opaline male stickleback with their blush-red bellies in the burn. Ebony male black grouse met in the early morning mists with their red eyebrows flaring to display to each other in irate bobbing, kicking, clucking ceremonies for their spectating groups of mild brown hens. The inland colony of black-headed gulls with lipstick bills reeling in screeching cacophony above the tussock swamp where they nested in their thousands provided a vibrant spectacle. The mink that escaped from a fur farm next to the village abattoir finished their raucous existence long ago. I remember the indignance of the minute cinnamon weasel I disturbed under a sheet of corrugated iron in the derelict glasshouses next to my home, a decrepit, forbidden realm of wonder. In the rank grasslands that grew there, grey partridge hid their nests of olive eggs and mating banded snails joined themselves together with love darts on dew-dripping grass stalks.

Wonders were always possible. When I opened my bedroom curtains early one summer morning, a tropical bird was feeding in our garden! Its face was black, its back ash-grey and its radiant breast a blushing Jaffa orange. I woke mum instantly, but when she opened her curtains and peered out it had gone. She never saw the male redstart.

Absorbed as I became by sheep in my teenage years, wildlife was only of notice if it killed mum's chickens. One of the dealers I used to work for told me that an eagle had been seen near Newbigging, not far from my home. I never saw it and wonder now if it was simply a buzzard that had attracted his uninformed attention. I cannot remember ever seeing even a single one of these now-ubiquitous mewing soarers over my land when I was young. The magpies, which likewise now abound, and ravens were also not present in my border days. The prospect of a pine marten or otter living in the landscape around our house was utterly implausible. They were too far away and too many gamekeepers with a lust to line their gibbets still lurked in the undergrowth.

CHAPTER TWO

Like an Aunt Choosing Cakes in a Tea Shop

T
O BECOME A ZOO KEEPER WAS MY EARLIEST WISH.
Mother recorded it in my baby book. It appeared
to me the most marvellous of professions.

The TV series *Animal Magic* was at the height of its fame
and I would curl up with my brother once home from
school to watch Johnny Morris in black and white bim-
bling his uniformed way through a working day, talking
to zoo beasts who talked back to him. Although neither
his lips nor theirs ever moved, we did not consider this
sinister or odd. The African zoo animal hospital series
Daktari featured a cross-eyed lion called Clarence who
had a genuine squint in real life and a chimpanzee who
rode on his back called Judy whose teeth were removed
to prevent her savage biting. Clarence's stand-in, Leo,
for the roaring scenes was impressive. Having been reg-
ularly beaten by his former owners as a cub, he could
only be filmed from behind very secure sets of bars as

his bite would without doubt have been much worse than his roar.

In the fictional books of Willard Price, Hal and Roger, the sons of the veteran animal dealer John Hunt, strode the globe in search of adventure, acquiring gigantic wild creatures to fill the cages, tanks and aviaries of zoos. *Look and Learn* magazine had articles about tough ranchers who stunned jaguars with shovels, leapt without care upon caimans and bound anacondas tight in a noble effort to 'bring 'em back alive'.

In a manner more sedate, like a great aunt cooing over cakes in a tea shop, David Attenborough quested for dragons on Komodo and picked his way through jungles to pluck out the yellow-headed picathartes, or rockfowl, for its return flight – willing or not – to the Zoological Society of London. Bald Desmond Morris championed vampire bats. When he fumblingly allowed one to flee in a warm studio on a live *Zoo Time* broadcast, it so terrified the cameramen that they fled the studio, hands high in a shrill shrieking herd.

Gerald Durrell was the doyen of them all. Bearded, jocular and rotund, he described with delight his adventures in the 1950s to collect the strangest of creatures for the zoos of the world. In pre-independence countries whose names are now near forgotten, he smoked out hollow trees, set nets across streams, crawled down into earthen lairs or paid hunters to keep and not kill, the creatures he desired. Along the way he danced with South American hookers, arranged animal deals with dictators, broke his ribs falling off Land Rovers and quaffed gin with the gigantically ebullient, alcoholic Fon of Bafut Achirimbi II. While giant anteaters crapped on him and he got bitten by poisonous snakes, Durrell's books engaged with amusing tales that warmed. As his thinking matured in the 1970s, he changed in rhetoric, if not always in practice, the way that zoos thought about themselves. Prior to his time, most were simply squalid versions of showmen's menageries run for facile entertainment and nothing more, but his contention that they could, and should, play a role of significance in saving threatened – and commonly unspectacular – species from extinction was almost entirely novel and his vision of them as purposeful was in its time near unique.

Carl Jones was one of his protégés. Laconically tall and Welsh, with a sinister ghoulish laugh, Carl took over a recovery programme for the native kestrel of Mauritius in 1979 at a time when the species was considered to be the world's rarest bird. Despite opinion concurring that his wards were doomed to extinction, he began an innovative programme of active management involving captive breeding and the provision of food support for the remaining wild birds. His plan was so successful that 333 kestrel chicks were produced between 1983 and 1993 and, although still rare, the species' prospects stabilised. Through using similar methods, he saved the dwindling echo parakeet and the island's gloriously pink pastel pigeon. He helped establish the Mauritian Wildlife Foundation and, with others, began a project to rebuild the ecology of the atolls that surround Mauritius and Rodrigues, such as Round Island, which had been destroyed by invasive goats and rabbits.

As a result of the forgoing, he is now very old.

Recently I met him in a skip in Llandovery.

I was driving back through the town centre on a muggy summer's evening when I spotted the stuffed head of a sable antelope in the front window of a small junk shop. As I passed, I thought about it and, having considered for around thirty seconds that although it was not British – my normal penchant for collecting stuffed creatures only applies to historic specimens of native sorts – I was highly unlikely to ever see another of its sort. I turned round.

Looking at the head with greater care in the shop window, I saw it was splendid. A gigantic, doe-eyed beast with badger stripes on its face, a fine erect mane and carunculated horns raised in a backswept arc above its head. I like wild antelope, goats, deer and sheep. They are lordly and I knew in that instant that although my cottage was small this beast simply had to come home. As it was mid evening and the shop was shut, I was considering a return visit when I noticed in the window a tiny card with an emergency contact number. As this was without doubt just such an occasion, I phoned the owner. She was prepared then and there to both reopen her shop for its sale and to offer a generous discount, so I nipped down to the local Co-op's Cashline to obtain her required fee.

At the side of the Co-op was a skip. As I passed it, I noticed that inside were a number of perfectly good wheely bins, which, although filled to their rims with detritus, were otherwise in good order. It was obvious that their rescue was required to save them for a useful function such as the storage of dry animal feedstuffs. It was a quiet night and no one seemed to mind when I drove my truck alongside, lowered the tail gate and climbed into the skip. They were heavier than I expected and the sliminess of their former contents, which had decomposed and oozed down their sides, made handling them on an uneven surface tricky. I had nevertheless emptied two and thrown them in the back of the truck and was reaching for a third when I heard a voice saying, 'Derek Gow, it is you! I knew it must be when I saw you climb into the skip.'

It was Carl.

He lived near the town and had stopped to purchase a bottle of wine. We spoke of this and that, of progress on projects we were both involved with and other matters of general natural history interest. He asked why I was so far from my natural habitat in Devon and, when I explained the saga of the sable, a strange glint came into his eye,

'You bastard,' he said. 'I wanted that but thought the price was a bit steep.'

When I smilingly explained that a discount had proved possible and that my uplift was imminent, he turned to get his wine without further delay. Now, it's a little-known fact that Carl is a collector of animal artefacts of the most competitive sort. He has stuffed giant tortoises in his house, which he uses as footstools, and an ossuary of skulls. In his mind, the sable was to set to join them. I acted swiftly. Relieving the last bin of some bricks, which were stuck to its side with dried slime, I turned the truck and took myself back to the shop. The lights were on, the owner inside and I had no more than seconds to leap from my vehicle, hand her the cash and turn before Carl was right beside me.

His speed of movement was extraordinary.

'You bastard,' he repeated with a hiss. 'It's mine.'

'Not any more,' I said, 'Too slow, but for the sake of nostalgic memory you can help me load it into the truck and I will send you a photo of it on my wall when I get home.'

With limited grace, he helped me to place it on the back seat before informing me that he had so positioned the horns as to ensure my impalement in the event of an emergency stop.

As I drove away waving, I could see his mouth moving.

'You bastard,' he said.

———

I left school when I was seventeen as I could see no prospect of my penchant for illustration resulting in any serious prospect of employment after a spell in art college. In truth, I was done with formal education and, with the exception of a miserable stint in a glum day-release polytechnic in a vain effort to acquire a basic mathematics degree, which I predictably failed, have never looked back.

Like an Aunt Choosing Cakes in a Tea Shop

Still enthralled by farm life but with with no prospect of being able to buy sufficient land to start up on my own, I did the next best thing and became a livestock auctioneer. My first job in the early 1980s was in an old market complex in Edinburgh with wrought-iron pens painted sky blue and ridged concrete floors to prevent cattle from slipping and breaking their legs. Wm. Bosomworth and Sons was cold and dark under its tall railway canopy of wire-reinforced roofing glass and corrugated sheet metal. Grim on even the warmest of days, in the years of its miserable function it had been a processing centre from life into death for millions of old, poor sheep. The sadness that its physical setting inspired pretty much permeated the people it employed and I left there very happily indeed when a full-time job as a principal auctioneer arose in my home town of Biggar.

Life was more fun in a homely environment. Store stock sales on Saturday, fat stock on Wednesday morning. Lots of trips in between to drum up custom from our farming clients, value livestock or to sleep after nice pub lunches at the side of a quiet forest road on a Friday afternoon. Even though its decline had not become clear, farming in our part of the world was constricting and times for my employers became tight. Commercial forestry had begun to consume many of the surrounding hill farms, and big progressive farmers were displacing the small. In the end, to justify my full-time employment, I was required to deliver the early morning meat orders from the abattoir attached to our cattle market daily to local schools.

This meant a start at 4.30am on a dull delivery round, which was only enlivened when parts of a pig embryo spat out by a malfunctioning mincer were discovered by a child, encased in the Cumberland coil he had nearly consumed. The fact that it was only pork of a slightly different age to the rest of the sausages' content placated no one, from the urchin who retched over his teacher's shoes to the Scottish Minister of Health. The shit hit the fan. There was lots of shouting and accusation. No one saw a funny side. The school contract, and my job along with it, terminated swiftly thereafter.

My time in zoos began when a friend, Dr Brian Thomson, offered me a seasonal job at Palacerigg Country Park on the outskirts of Cumbernauld in central Scotland looking after the flocks and herds of rare-breed domestic livestock it maintained. Within a year, when a staff member left, he asked if I wished to manage its captive collection of wild animals.

As a reality it was a rude awakening. The infrastructure I inherited was old. Heavy wire cages of rusting rectangles contained a pair of old wolves (one of which was blind), some wildcats of a similar vintage, a red fox with half a tail, deer of various sorts, a limping mink whose leg had been crushed in a fen trap, a short-eared owl with a broken wing and a rabbit colony, which contracted myxomatosis within days of my beginning and suicidally expired. This assembly of the blind, broken and diseased was an uninspiring inception. Whatever tint of rose naivety applied to these creatures in my charge, an ark sailing with purposeful intent they sure as hell were not. Brian was unconcerned about this present, which was good given it was awful. His was a vision of newly designed, enriched living spaces filled with trees, pools and bushes of delightfully sculpted environments, which represented the habitats of the creatures they contained. He wanted to keep and breed species of nature conservation importance and bless it with sound, educational purpose.

It all seemed so unrealistic.

As a bribe to entice me into accepting responsibility for the required renovation, Brian arranged that I should attend a summer school at Durrell's zoo on Jersey. While in large part this would focus on the experience gained caring for internationally threatened species such as golden lion tamarins, Jamaican hutias or coneys, volcano rabbits and Rodrigues fruit bats, some of the principles involved and lessons learnt might have an application for us. The development of this course at the then Jersey Wildlife Preservation Trust had been entrusted by Durrell to one Simon Hicks in 1977 and remained unique in its sort when I attended in 1990. Slim with a sort of wiry military erectness and a mop of curly brown hair, Simon was and still is a hugely enthusiastic man. He likes to listen, think, understand and ask lots of detailed pointy questions that hurt the inside of your head. It was his contention that a two-week course, which brought young people working in conservation the world over to drink cocktails in Jersey's least-exclusive nightclubs, blag entrance thereafter to its Irish bars to sing folk songs and then fill the municipal flower borders of St Helier brimful to the top with vomit, would do wonders for the plight of threatened species.

While I can't remember a thing we were taught formally, I can keenly recall how together in thinking as a group we were in the depth of the passion we all had for our fond subjects of choice, be they slimy, furry or feathered, and the issues surrounding their salvation which impacted us all. I met people from countries without all that in Britain we thought normal – photocopiers, carpets, personal assistants, Biodiversity Action Plans and governments or opponents who, however much you annoyed them, were very unlikely to intentionally kill you. Those folk, and many of the same ilk I have encountered the world over in times since, who have committed themselves utterly to the salvation of wild creatures, do so quite commonly to their own individual detriment. Dissolute relationships, poverty, despair and all its following ills can follow the dedicated charismatic. It's not always an easy life route to traverse.

Brian's thinking was progressive and, as more resources were wrung from its tight-fisted owners, new opportunities arose. As his wife was Norwegian, he had a fondness for Scandinavian wildlife and importing the white storks he had organised from Sweden meant designing new aviaries, building nests and organising quarantine, diets and healthcare. Once these glorious birds were in place, could we get them to breed? Well, yes was the answer and, after that, new breeding groups of wolves, then reindeer, then wild horses. Year on year, the challenges we were given or set for ourselves afforded a delight of detail. Conversations with continental zoo directors about the care of chamois, the husbandry of mountain hares or the diets of capercaillies were absorbing. I learnt so much! The collection of big horned sheep the size of donkeys whose head-butting blows knocked their keepers at Whipsnade back out through a pen door when they tried to catch them for transport to a paddock on our Scottish moor was a lesson as well. These physically powerful creatures were as it turned out surprisingly susceptible to stress; harass them for too long and heart failures would simply drop them dead. Best to act fast, then move them calmly back home in dark, padded crates. I jumped in to hold them and to help without care.

It was all just so exciting!

My wooden house in the forest was shaped like a gigantic isosceles triangle. As this was the nearest staff dwelling to the car park, my girlfriend Shona and I were first to receive any persons delivering wild waifs. While for some of these creatures redemption was possible, others were just too far gone for anything other than their last rites. The most memorable was a still-breathing rabbit that had been so obviously squashed by a car that tread marks were indented into its flattened rear end. How it survived for even a short time was beyond comprehension and, despite the entreaties of its finders, any prospect of reinflation to save its existence was quite simply impractical. A sprite-like red squirrel hand-reared by a friend escaped from its cage in the front room of my house and bored its way into a horsehair sofa I had been given by my mum. When I went to work during the day it emerged and gnawed through the doors of the kitchen cupboards to steal raisins, nuts, breakfast cereal and pasta. It cached its plunder all over the house. Years after it had been transferred to a breeding programme in Norfolk, its food stores under the bath, at the back of an old sock drawer and inside the face mantle of a grandfather clock were still being unearthed.

I reared deer of many sorts. Tiny, delicate roe in dappled coats on spindle legs, larger fallow deer and one tiny reindeer calf whose mother had no milk and which looked like it had been constructed entirely out of dark brown Velcro. A wild boar piglet called Truffle, which, although hypothermic and blue, had rapidly revived when installed under a glowing red heat lamp, lived in the house for some months. She was a treasure. A tiny, trotting Disney diva come alive with great dark eyelashes and long cream flank stripes. When she was very small, she was fed watered-down Carnation milk from a tiny kitten bottle with a minute rubber nipple. As she got bigger, this became a human baby bottle and then a full Pepsi container with a large screw-on lamb teat. She had a warm stable pen during the day but walked home every night to the house across the zoo yard. She would race to the garden gate and then to the front door and squeal with increasing impatience as these were unlocked. Once inside, it was straight to the front room for her blue and red flowery beanbag, which she would drag across the floor, place in front of the wood burner and stand, grumblingly nudging with her snout until the first flames drew and we stepped back. At that point she would flip the beanbag right under the stove's windowpane and flop in to toast first one and then her other stripped flank. On occasion, when the heat was fierce and the smell of singed pig hair filled the room, you would try to move her from her chosen spot. She opposed this prospect fiercely. Although tiny, her tusks were sharp and I still have a long straight scar up my right-hand thumb that testifies to her irritation at being moved from her prime roasting position. When she got bigger, she did not want to sleep on the beanbag anymore and the sofa became her abode of choice. One day, having evicted her grumbling before nodding off on my own, I woke aghast to realise I had lost all sensation down one side of my torso. Convinced in that moment that I had had a stroke and was dwindling fast, I opened a rheumy eye to see a great fat russet pig's face directly opposite my own. Its eyes were tight shut but when it snored, its lips moved and dribbled just sufficiently enough to create a large wet patch on my tummy.

That did it. Fighting pigs for the use of my own sofa was and is never going to be a tolerable pursuit. Truffle went back to the zoo, to a mate and the comfort of a large woodland paddock where in time she had many stripy offspring.

Still the house filled up. There were fox cubs that leapt and sprang in strokes of flowing, fiery grace from bookcase tops to windowsills to sofa backs to you, but my favourite by far, and the most characterful of all, was a baby brown hare. Its mother had been hit by a silage cutter and the farmer's wife who found her mangled corpse with the leveret alongside was utterly distraught. I had tiny kitten teats for feeding an appropriate milk substitute, but simply could not get him to suckle or to feed. Nothing made much difference and hourly he faded. In desperation, I cut a large turf from the garden filled with fresh sweet grass, dandelions and daisies and placed this in the bottom of the warm dark box he was occupying in my bedroom. An hour later he was mowing, then lapping milk from a small dish I placed in the middle.

I changed his turf daily and within a week he was out of the box and exploring the house. As he got bigger, he became more confident and would race down the hall, perform a perfect pirouette in the kitchen and, at speed, supersonically leap from the front room door onto the beleaguered squirrel sofa in one bound. There he would sit sedately and calmly groom his coat and ears with great care.

My cat at the time hated him. Always ambivalent about house fauna, slow ginger Tigger had taken the hump when the hare started to stand on its hind legs to punch him in the face. Hissing did not help – the hare was too damn confident – and Tigger, at least while I watched firmly, would leap to table height to get out of his way. One night, stretched out in front of the wood burner while the hare slept across my chest, a knock on the door caused me to lift him down to attend to the query quickly and return. I was only gone moments but it was all the time Tigger needed to kill.

I wept when I dug the hare's grave in the garden and buried him, with his perfect brown body soft, warm and still curled into his blanket.

———

In 1991 I was sent to acquire a female European bison from a small zoo in Wales that was closing. We wanted to established a breeding herd of this species, which, although once widespread in mainland Europe, had been greatly reduced in its range by the time of the First World War. At the outbreak of conflict, the largest herd remaining in the world lived free in the one hundred and fifty square kilometre range of the Białowieża Forest in Poland. When the invading German troops reached and occupied the forest, it is believed that they killed over six hundred despite the protestations of some of their officers about the rarity of the species. When they retreated at the war's end, only nine remained. The last wild bison in Poland was killed in 1921 and the last in the Caucuses in 1927. Fewer than fifty remained at that time in the world, all of which were in zoos. A stud book was founded to help conserve their genetic variation in 1923 and, although many of those bred in captivity were killed in the Second World War, the species now numbers around seven thousand individuals and has been reintroduced as a free-roaming species into many areas of its former range.

As further reintroductions were in planning and we were keen to participate with the conservation breeding programme for this species, we drove from central Scotland in a flatbed transit van with a brand-new cattle trailer attached to collect the cow in Wales.

We loaded her without issue from a cattle pen where she was used to being fed before driving cross country to what was then called Port Lympne Zoo in Kent to collect her mate. Mike Lockyer, its collection manager who had identified a suitable surplus bull from their herd, was both physically and metaphorically a huge character with a vast waxed beard and moustache flowing out in fine Raleighian form. At the helm of an Elizabethan man-o'-war wielding a cutlass he would not have been out of place. Mike had caught herds of animals in Africa for the Chipperfields when their circus and safari park empire was at its zenith by jumping out of bouncing Land Rovers to grab and then bind them with ropes. He had supplied dart guns for Idi Amin's personal firearms collection, bribed the widows of the Africans who died catching buffalo bulls with cornflakes to keep them silent, and generally seduced and caroused his way around the continent in a manner that would have left Sir Walter green with envy.

The bull bison, once loaded, was not happy. Although near adult size at around the age of four, he was still somewhat frisky. He did not like being separated by a sheet metal partition from the cow he smelt in front. He did not like leaving his herd. He did not like being in a trailer or the hangover he was nursing from the immobilising drugs

– and he definitely did not like us. While I prepared with my co-driver, Fraser, from the Western Isles to settle down for the long drive home from Kent, the bull, snorting and head high, prepared for something else. We had just reached the orbital of the M25 when the first boom sounded. The trailer swung and our towing van shuddered. Others followed in a series of shocks and, although we rationalised that it must be the bull bashing the sides with his short-horned head, we had no idea what impact this was having. We stopped in the services at Clacket Lane and got out to see. The bull stopped bashing and Fraser, who was small, climbed up to cautiously to open the trailer's top vent for a peek. As he looked in and turned to speak, whatever he was going to say was supplanted by 'ohhhhhhh, fuck' as the bull hit the trailer and Fraser flew high in an arc to land flat on the tarmac. As he lay there groaning, passers-by in neat clothes, some holding the hands of tiny children, looked on aghast. The bull resumed bashing and the trailer, which should have been static, leapt up and down. The passers-by picked up their pace or ran.

I called Mike and because he did not want the surplus bull back, he offered to come up and shoot it. In the car park. In thirty minutes' time. I considered his offer for a few moments and, unsure as to whether it was a studied exercise in sarcasm or not, declined as I could think of no possible cover up explanation for an action of this sort in what was arguably one of the most visible venues in Britain. Short of driving to the gates of Buckingham Palace, ringing the doorbell and then blowing the bull away with a blunderbuss in front of the Queen holding Prince William's hand while he cuddled a baby corgi, I can imagine still no location more obvious for debacle.

I pulled Fraser up and we hit the road north again – faster. For a time, this seemed to work. The bull stopped. We hoped he had gone to sleep. Around Manchester he revived and the booming and swaying began again in sequence. He did not cease when we drew into a fuel station to fill up and had to weakly explain our plight to a perturbed shop assistant who had never before witnessed a cattle trailer with hiccups. On up the road we went until near Penrith the sounds in the back changed to the kind of popping noise anyone can make if they just sit still and open their mouth quickly with their lips rounded. Only louder. Much louder.

I looked in the wing mirror and, to my horror, saw that the bull was punching its horn through the by-now pitted sides of the trailer. Although we knew that nothing that gigantic was ever going to come out through the sort of modest holes he was creating, the prospect that the force of the horn's powerful withdrawal would pull the panel rivets free and, by doing so, open the trailer's side to the world was petrifying. With all the haste we could muster, we sped forward on the journey from hell that was not near done. We got to the park late at night, where a tractor with lights on was waiting. Without further ado, we reversed up to the new bison pen gate and opened the door.

The bloody bull exhausted by his long day of vigorous activity was sleeping like a baby on his straw bed inside.

Like an Aunt Choosing Cakes in a Tea Shop

Nothing we could do would persuade him to exit the space that he now considered his home. We left him where he was to come out when he so desired and set the calm cow free in the morning to join him.

The more I witnessed its characters, the more I read, the more sites I visited and the more informed I became, the more acutely obvious it was that zoos were schizophrenic organisations full to the brim with the oddest of folk. Animal dealers with Rasputin-like morals, the purchasers of penguins that whalers had wounded, the wrestlers of dancing bears, circus folk who danced naked on elephants' backs, people retired from insurance broking who just wanted to own a zoo in their years of decline. Nothing much was needed other than a willingness to say you could and the cash to seal a purchase. Crooks, cranks and weirdos galore did so with glee and with gusto.

Some, like good, gruff Frank Wheeler, who was able and ran London Zoo's collection of bewhiskered small mammals in the Clore Pavilion, were diseased. He was one of a tiny group of folk worldwide unfortunate enough to have ever been bitten by a sloth.

'How did it happen?' you would ask.

'Very slowly,' he'd reply.

The net end result of his condition was a nasty psoriatic skin, which, on the worst of days, would leak clear fluid from patches of red inflammation. Frank was a friendly soul, always keen to chat to any pals old and new, who visited him in his underground lair. On bad days his fingers would itch so badly he would scratch them with his teeth. This action, which caused a light snowfall of skin flakes, was always viewed by informed visitors with great apprehension when they accepted their coffee or cakes from his moist hands.

Dan Hadley was an old colonial sort with erect startled eyebrows. In his youth he had been a superintendent in the Zambian police before independence led him to manage several small zoos in that country. He was swart, brusque and firm. When I first met him in the early 1990s, he walked by turning his body from side to side. Like a shiny toy robot. One leg forward, then a heave to move the next. On enquiry, it turned out that this form of locomotion was the result of a faux pas from a time when early black-and-white cine footage of a charging lion required one to sit in a small trailer behind a Land Rover with a camera. A dead zebra tied to the trailer with a rope was the lure. While, as a system, he said it generally worked well, on the day of his downfall, the lions were tired, lazy and languid. However close the zebra was dragged, they did not bother to blink. After several efforts, his driver had stopped the Land Rover to discuss, with Dan still sitting in the trailer, the viability of other options. Suddenly he screamed 'Lion!' and leapt back to his seat. Dan turned just fast enough to see a maned male race past the zebra and head straight for him before the shock of the

Land Rover's fast-forward start catapulted him out of the trailer and into its mouth.

It crushed his pelvis and shook him like a rag doll and, although the driver saved his life by swiftly turning and running it over, his form of personal locomotion was altered ever after.

His eyebrows remained bolt upright for the rest of his life.

Alf Robertson, who ran a small zoo in Dundee, looked like an old bald mate of Thorin Oakenshield who had swallowed Rapunzel's grandmother whole and then forgotten to pick out his teeth. Any impression that this might afford of infirmity is wrong. Alf was hard as hell. In the days before dart guns, he was famous amongst the zoo fraternity for wrestling every creature he had ever possessed to the ground, using a blunt implement to begin with but then his hands if all else failed. He had a phrase he used with flair when his staff, injured grievously in previous conflicts, refused to help him in later life: 'I will sort the fucking fucker out myself.'

Jo Jo was a large macaque who had done time with his previous owner in the 1950s British merchant marine. He had lived through some hard times as a youngster and witnessed many things that no monkey ever should. Harrowing experience had, however, taught him to drink rum, smoke cigarettes and shake hands. When he came to Alf's zoo, he was young. When Alf had to move him from his old cage to nice new enclosure in a sunny walled garden, he was late middle aged and corpulent. He liked his old home although it was small, cramped and smelly. He did not want to leave.

Some effort had been put into baiting him into traps, which he had avoided with disdain. Drugs had been tried in his food. Jo Jo had simply ignored the treated items or, depending on his mood, thrown them back with some poo at his captors. The old zoo was being demolished; he had to be relocated, pure and simple. On the appointed day, Alf came into work early. He walked into the staff room and stated that the day to move Jo Jo had arrived. When they asked how, Alf said that they would net him and then, using their collective strength, pin him to the ground before crating him. His staff said 'fuck off, he's dangerous'. Alf called them 'wimmin' and, grabbing the nearest net, marched off towards the primate's cage in fine gladiatorial fashion. By the time his staff gathered in haste to spectate, he had already entered the cage.

Jo Jo had jumped to the floor and was standing erect with his lips drawn back and huge ivory incisors exposed. Height-wise and in terms of weight they were near evenly matched. Alf screamed at Jo Jo, Jo Jo screamed at Alf, and the battle that both knew was inevitable began.

Jo Jo caught the net on Alf's first swing and with ease snapped its head clean off. Alf looked at the pole, Jo Jo looked at the net before, in synchrony, each threw each part at each other. In a screaming mass of fur, blood and faeces, they engaged. One moment rolling on the floor trying to gain advantage, another bashing bodies against walls. Alf was on top but Jo was ripping his clothes asunder. Jo Jo kicked Alf in the face before being grasped in the firmest of wrestling holds. Both bit each other. Swearing, grunting, renting, beard fragments, white hair, garments in shreds filled the air.

In the end it was Alf who lasted longest, marginally.

By the time all movement inside had ceased and the gawping staff entered the cage to drag the assailants' bodies apart, neither was mobile. Jo Jo was transported with haste via the vet to his new home, while Alf went to Ninewells Hospital for significant stitching and a blood transfusion. Both lived on in the zoo for many years and would on occasion pose as old adversaries do for photographs in the media, sharing a smoke while smiling at each other and shaking hands in the sun.

When Alf retired, his assistant George Reid took over. Eloquent and loquacious, George was an elegant charmer. He would have made a splendid courtier, adored by queens, and though he never wore a ruff to the best of anyone's knowledge, it would have suited him well. George liked to play with animals. Unfortunately, as the scars on his hands and face testified, they did not always want to amuse themselves with him.

One Sunday morning he was walking through the zoo. Its new location in what had been the old walled garden of Camperdown House was pleasant with large enclosures for its collection of British wild animals and birds. It was spring and the flower borders were aglow with colour. Wild red squirrels scampered across the paths and the earliest of nesting birds sang sweetly in the tall lime trees.

As he passed the badger enclosure surrounded by a low brick wall – badgers, although good climbers, are not great jumpers and it does not take much of a height to contain them – George observed that the male was out foraging for worms in the soil. He knew it was a beast of mixed temper but, on that day imbued by the spirit of nature's amicability, without thinking he stopped to tickle it under the chin. In response to this friendly gesture, the badger bit him firmly in the soft of his hand between his thumb and his first finger.

Badgers lock their jaws shut. This adaptation gives them immense crushing power, which, once exercised, can require gelignite to relax. George at that point had none. The badger refused to let go, despite being tickled even more desperately by George with his other hand. Despite his calls for staff help, none appeared and, just when he had come to the conclusion that the only solution was to lift the badger over the wall with one hand in the hope that he could then drag it like a furry millstone back to the work yard for aid, another problem arose.

Woken by the commotion above the female badger emerged from her nest. Irritable and sleepy, she moved to examine what her mate had caught. George pushed her away with his other hand. She bit it in near exactly the same position as the male, locked her jaws as well and, as badgers do with something they desire, began to then back herself steadily towards her underground sett entrance. The male followed her example and George, with arms crossed and strength failing, became inextricably entwined in a by now desperate juddering tussle to prevent himself from a subterranean end.

Just when it seemed that the experience could get no worse, it did.

Out of the corner of his eye, George spied two elderly ladies whose delight it was to perambulate the gardens in the early morning, arm in arm, approaching him slowly. With his final remaining strength, he turned to face them, arms crossed behind, badgers dangling and summoning with the last grace he had, said, 'Good morning.'

'Good morning,' they replied in unison. 'It's a lovely day.'

'Yes,' said George, jerking. 'Yes, it is.'

'The gardens look marvellous, Mr Reid.' they said.

'Yes,' he said, 'yes they do.'

'Do you know, we saw a squirrel?' they said.

'That's nice.' said George, as the male badger crunched harder.

'Well, we must be moving on now, Mr Reid. We know you are a busy man.'

'Thank you very much,' said George.

As his eyes closed in relief, with his badgers still tugging, George overheard heard them say as they moved away:

'Isn't it lovely to see Mr Reid playing with his badgers!'

———

While zoos in my time afforded spectacles of every sort in purpose and intent, they were hollow. As far as British wildlife was concerned, few saw much potential in their cultivation. Things like water voles were small and fiddly. Big zoos did not understand the complications of their care, the requirement to manage them in numbers like an annual crop or how poor an exhibit they presented when compared with a zebra. When an earnest, beardy chum of mine, Simon Wakefield, who was then a young zoologist, took the first rare barberry carpet moth that Natural England gave him to start a breeding programme to show his director, John Knowles, John took the jam jar it was in, leant back in his swivel chair, sucked air through his teeth and told him that never before in his life had he ever been so underwhelmed. While admittedly barberry carpets are very dull and John had a bloody good sense of humour, his response at that time would have been pretty symptomatic of the response given to any proposal to assist the conservation of small British species.

How, after all, could you compare a pine marten's stage performance as an exhibit to a Siberian tiger – or a giraffe or a gorilla. If that's your benchmark to begin with, they will never measure up. To be fair, John was uniquely a supporter of one of the first projects to use the prehistoric Przewalski's horse, with its erect black mane and zebra-striped hind legs, for conservation grazing at Eelmoor Marsh in Hampshire where, along with High-land cattle, they have done a great job of gardening this Site of Special Scientific Interest (SSSI) habitat for the general benefit of its smaller species. Nature conservation for the old directors was not, however, generally based upon ventures of that sort. They much preferred to build gigantic penguin displays in the shape of a leaping sperm whale, designed by international architects based in the Seychelles at costs that would have confounded Onassis. Quite what, how or why it contributed to conservation, no one ever understood.

Although this has now changed, with many of the best doing credible work both in Britain and abroad, in my time leaders would lean on lecterns and espouse to audiences of the completely converted home-spun credentials for the preservation of glamorous species whose futures they could not possibly hope to influence. Sure, animals moved between zoos for breeding programmes with fancy titles but did this not have more to do with supplying creatures for display rather than anything else? It was very obvious to any who thought hard that releasing tigers, which had been captive bred long enough to develop hip dysplasia and lisps, back into parts of India that had become housing estates was not going to prove practicable.

While a few actions of worth were undertaken, these were generally small in scale and very limited to the interests of the individuals involved. If you found someone with their hands on the purse strings who was keen or knowledgeable or cared, then good things might happen. It's always individuals, single people, who change things. Take Iain Valentine, a gangling russet-haired chap with large teeth who, had he not had a deep interest in animals, would have been designed by nature for comedic, vaudeville theatre. Iain was, and still is, brilliant. When the first beaver application to release the species in Knapdale was turned down, it was he who, as the collection director of the Royal Zoological Society of Scotland, picked the dead project up and blew on its embers. His intelligence and rounded experience of Scottish field sports, farming, its shimmying political back drop and what had been accomplished with beavers elsewhere in Europe made him the best man for the job by far. He understood the needs and wants of the funders and other project partners and played all the forgoing together at times in masterful symphony. Without Iain, it's very doubtful that any of the other organisations, which now claim credit for the project, would ever have accomplished much. He ensured the project's rebirth.

I moved from Palacerigg when I felt that I had achieved much of what I could to begin work on a fresh project in the New Forest in 1994. I wanted a change and, when I was approached by the Sea Life Centre group of aquaria to develop their first mainland wildlife attraction in a warm, scented pine wood near Ashurst on the forest's edge, I said yes.

My new boss, Laura Wade, was clever and garrulous – I have never worked for anyone I respected more. We argued like hell. She told me to do things that seemed idiotic and I told her it could not be. She showed me I was wrong time and again. In the end I came to believe in her pretty much utterly. To inform our thinking, we travelled to the best of the most wonderful zoos in Europe where smart people had ideas well advanced beyond ours. To the otter centre at Hankensbüttel near Hannover, to Royal

Burgers' Zoo near Arnhem and to view the superb exhibits of large mammals in the Bavarian Forest National Park, whose enclosures were so extensive that tree saplings grew in their woodland paddocks. In all, smart designers had wrought living spaces very large or small to specifically cater for the behavioural requirements of a single species. No general thinking of just a big cage with branches to hold a monkey or a lemur or a fossa or anything else that climbed and came along. No big paddock homogeneity where a giraffe or a camel or a yak or a wildebeest might all equally be stuffed without a lot of grumbling. Detailed understanding and design for the needs of each subject and its requirements. No one size fits all.

At the excellent Alpenzoo Innsbruck, cunning tricks such as slanted mirrors above sleeping boxes that you viewed through slits, one-way viewing glass or hidden cameras in nest boxes were all installed to ensure that you could see the occupants even if they desired not to see you. Dippers flew underwater in a vast aquarium through and past rocks to consume trout eggs scattered at random at their base. Reptiles basked in angular rocky enclosures in the Austrian autumn beneath cunningly hidden heat lamps in the roofs of their outdoor vivariums. Most exhibits were designed to mimic the living space that the animals would naturally occupy but, even with this simple premise, imagination within these intelligent institutions flourished.

Beech martens are a near relative of our native pine marten in Britain. Unlike ours, they have white chest bibs and bellies, more weasel-like convex heads and a paler, almost peach/russet coat. If you have ever owned or stayed in a rural gîte in France, it is these *foina* that raid your chickens at night, build pyramids with their own shit in your roof and, when they feel frisky on a warm summer night, produce a bewilderingly broad range of poltergeist type sounds to terrify your slumbering kids.

They love the upper stories of human dwellings – the roofs and lofts – which they access by squeezing up under the space where their beams meet the top wall. One lovely old German farmhouse designed in its lower stories as a visitor centre during the day with stuffed creatures, aquaria and educational panels had converted its loft into the most amazing night-time exhibit. Tame baby beech martens, which were commonly rescued as orphans when their parents were either shot or otherwise killed, were housed in large aviaries outside. Although they slept during the day, at night-time they emerged and entered the loft through a tunnel. The space inside, which contained partially opened old wooden chests with old clothes peeking out, piles of newspapers tied up with string, a gramophone with its trumpet still erect and discarded old kids' toys, had been specifically designed to contain them. To view their antics, you walked up a rickety set of outside stairs before entering a dark alley along the side of the building into which were set large glass viewing windows. The guide that accompanied you shone a red lensed beam through into the area beyond where you would realise, as your eyes accustomed to its light, that a rocking chair was moving. Suddenly a brown streak would fly across the floor, and then another, as the beech martens caught and tumbled in an ecstasy of play. Raisins, fruit, dead mice and tiny dog biscuits hidden throughout the exhibit during the day would fuel their agile explorations. It was an utterly exquisite and enchanting insight into the world of a creature most people would only ever see lying dead, dirt splattered, crushed and bloody on the edge of a road.

Our New Forest project was to be a new kind of zoo called Nature Quest, which Laura told us would be brilliant. We would build it like the very best we had seen. The company only wished to display, as a unique selling product, wildlife species that were to be found in Britain in the same thematic manner as their aquarium chains. This guild incorporated species past, present or introduced to the landscape by people, such as fallow deer or the smoke-blue edible dormice with dark mascaraed eyes. Released in 1902 by the Victorian naturalist Lord Rothschild, these dormice still inhabit the loft space of weary householders near Tring to this day where their gnawing abilities through the sides of biscuit tins in kitchens are legendary.

No other organisation or individual had tried to do this before. Many of the species we wished to work with were protected by law and the authorities who governed their existence would not simply allow us to display them only to make money. Even common species such as moles or wild rabbits were unknown captive subjects and, other than the broad range of deer types or wild boar, this situation applied to much else. Very little was known about the husbandry and care of British wildlife. While catching some specimens to form breeding stocks would obviously be essential, wild animals are commonly secretive and would make poor exhibits in any case. It became clear as we planned that the more settled off-spring of wild individuals would be much more likely to make better exhibits, and so we built a breeding facility that would enable us to produce these in numbers, with spare capacity for several colonies or pairs of each.

We were well funded, and joiners and fencing contractors were hired in numbers to make our plans a reality. Large circular tin pens four feet high and thirty feet in diameter were erected around landscapes redeveloped into small wetlands with concrete swimming pools for water voles or shrews. Tall woodland aviaries were built for pine martens and wildcats. A system of large holding tanks contained our fish stock from gigantic green carp with hard armoured scales to silver swift shoals of bitterling and dark lurking pike. Reptiles came from the New Forest keeper Martin Noble's own personal stash. Tall fences everywhere retained different deer herds, wolves and sounders of wild boar. Internal rooms filled to the brim with tanks, small cages and aviaries were constructed for tiny mammals, insects, amphibians and birds. Creatures of all sorts began to flow in. Some from wild trapping, zoos, aquariums or official breeding programmes. Others from wildlife rescue centres when they became too tame or were too disabled or injured to release, proved, like Peanut, the cuddliest fox in the world who threw herself shrieking into our outstretched arms for a hug, to be the best display pals ever.

We built the above on time within a year. From background research through creation to a well-advertised opening when eager visitors flowed in. Exhaustion and exhilaration combined.

Although the concept of displaying fascinatingly tiny creatures such as water shrews, which are black on top and white below and have large bristles on their hind feet to propel them swiftly when they swim underwater, was simple, its reality was complex. Intricate exhibits designed to delight in their first days where they could be seen underwater alongside the fish, lampreys or crayfish, which naturally share their world, or on land with appropriate rocks, roots, plants and insects, were hard to maintain. Small things like to hide. They like to eat the other things that you wish them to coexist with or to avoid being eaten by others that are larger than them. Learning was all. Did you know that a male stoat can have sex with a baby female when she is blind and naked in her mother's nest and that she will then naturally store the eggs he fertilises and give birth to live young in the following year? Were you aware that Scilly shrews will live quite happily as complete family groups and in doing so form linked caravans by holding on to each other's tails around their rocky exhibits? Who would have thought that tiny glow-worm larvae, which predate equally tiny molluscs naturally, would quite cheerfully survive on a diet of liquidised large snails? Or that the presence of hyper-dominant male common dormice placed in adjoining systems of aviaries would so intimidate more timid male neighbours that no babies would result?

We absorbed these lessons and adapted with little time to do so.

It had become ever more obvious that our expanding resource of captive wild creatures, designed initially to simply supply our displays, was of worth for much more. We became involved in research project after project with the best of universities studying the behaviours and lifestyles of the little known. Would water voles avoid mink shit, or fox, rat or wolf? What diameter of radio collar would fit round their necks, and could captive-bred juveniles born from wild-caught adults ever be expected to live free and successfully in the wild? Study after study led to better understanding. Our breeding programmes started to show results and, in the end, there were only a handful of land mammals – the hares, common and pygmy shrews, weasels and moles – that we were never able to encourage to reproduce. Alongside these, we bred many others, such as critically endangered shiny black field crickets, Duke of Burgundy fritillaries and swallowtail butterflies, natterjack toads, tadpole shrimps, sand lizards and more.

In parts by design and default, we had birthed a Durrellian model, which was tangibly capable of assisting the recovery of a whole host of threatened native creatures. Organisation after organisation, other zoos including

Jersey, producers of wildlife documentaries, wildlife celebrities and government bodies came to see and were impressed. Many offered further and various support. At that time, a well-established relationship with the government's nature conservation body, English Nature, was possible, and we bred considerable numbers of common dormice for their species recovery programme. While this rising whirl of enthusiasm pleased the company's PR folk quite a lot, it did not make the commercial mandarins smile. They were growing tired of a product that required complex adjustment. They wanted simplicity. A known cost for everything. Clear dividends for shareholders. A simple formula that could be replicated, run out at quantifiable cost, repeated and sold. Their focus was changing too. At Sea Life centres, visitors had had enough of haddocks, and their herring shoals in circular tanks were no longer spreading happiness. More exotic candidates were desired. Tropical fish in coral reefs were the future worldwide for them, alongside waxwork museums and theme parks full of rollercoaster rides. Odd projects the likes of ours were just weird and awkward.

Those of us responsible for Nature Quest's creation wished it to continue. To become the very best that it could be. To fulfil the promise that we were sure to our marrow it had. I, on my own by then as Laura had been seconded away, approached a conservation charity called Tusk Force who, at that time, was raising funds very successfully for overseas nature conservations works. They offered to help but no satisfactory deal could be done. Dr Lee Durrell and Simon Hicks came from Jersey and were both impressed. As they wanted to form a UK centre at that time for the then called Jersey Wildlife Preservation Trust, they felt that Nature Quest would be ideal. They took a proposal to buy it to their board, who said no. Other interested parties came and went. We were young and naïve and had no money. As it turned out, we could have purchased it for very little.

In the end it was sold swiftly to the first operator of a commercial zoo who showed serious intent. Roger Heap was a tough, brusque, northern businessman who, having made his money in other ventures, indulged his passion for wildlife in later life. He sacked me on the day he took over along with a number of my co-workers as he felt sure that his vision of what was required was not mine. On the same day, he then offered me his help to move forward and has done so ever since. His advice has always been sound and I respect him a lot. Many of the most important breeding groups of animals we had established over years he did not want, and we agreed that if I could rehome them they could remain intact. As the closure of the Nature Quest project attracted some media attention, I was rapidly approached by several individuals who wanted to help. Kind Sarah Bridger from Fitzhall Lodge estate in West Sussex came with her mum to offer their land and buildings for housing any creatures we might wish to rehome for as long as it took to assure their future. Laura and some other senior work colleagues left their jobs in disappointment at the company's stance and helped both with money and with labour when they could.

Ken West, a wealthy elderly businessman whose lifelong passion was nature but who had built a private fortune by providing airport security systems in Africa and the Middle East, asked me what I wanted to do. When I explained that my intention was to stick with the idea of creating an independent wildlife centre where native species of threatened wildlife could be bred in captivity for release back into areas of suitable habitat in time, he immediately said that he would back me, help buy another site, which could be redeveloped for this purpose, and start again. Ken, silver haired, tall and dignified, was endlessly kind and supportive in the darkest of times. He died in 2017, having committed much of his own money in trying to make this vision reality. Without him, my personal dream of accomplishing anything much would have expired with Nature Quest's end in 1998. I owe him perhaps everything now.

I kept the important species. The common dormice, water shrews and voles, the beavers that we had recently acquired from Poland, the pine martens, wildcats and some other tame creatures that I thought might be useful. I spent all the redundancy money I received, got a part-time gardening job with a friend and slowly, in my spare time, started to move the animals from the Hampshire site over into new enclosures in Sussex. It was an excruciatingly slow process. Friends helped in their spare time but my every waking moment was consumed by building and finishing cages. Day after day. In the rain and the snow. On cold dark evenings and, as the seasons changed, warm misty mornings. Feeding animals, sorting problems. Sarah Bridger was keen that a new centre should be built on her land but no money could be found to undertake this task. Other sites were proposed by Ken or a small working group we had formed called the NUArk Trust. I visited many sad old zoos for sale whose infrastructure with their owners' deaths was dissolving into the soil. Most, as prior planning permission had been given, would become housing development sites well outside our purchasing power. We looked at a farm park at Cholderton in Hampshire whose owners wanted to sell up, a bird garden near Flimwell in Sussex where ex-partners in an hate-filled dispute were parting ways, garden centres and other possible options. Meetings were held with the Forestry Commission to discuss other sites.

A year came and went with the animals, but any plan or prospect for forward movement remained elusive. I was approached by two retired foresters who wished my assistance with a simpler wildlife project in a woodland location in the Forest of Blean in Kent. Although the site's infrastructure had been levelled – its former use was a small petting zoo – and it had nothing much more to recommend it than a single tap on a standpipe, it offered a start. I left the group I was working with (which was still being supported by Ken) in frustration at its lack of progress and, in doing so, took most of the animals. It was an unhappy parting and, though some of us who were close still speak on occasion, it made nobody happy at the time.

The Wildwood Centre was a low-budget affair, which, although it kept alive to an extent the vision I wished to pursue, was hard work. More physical construction ensued as we created underground houses for badgers and mimic barns for bats, rats and owls. There were many large outdoor enclosures for wolves, wild boar, deer, birds and beavers. In time, Ken, when NUArk dissolved, returned to assist its development. Although I personally got on well with him and another of the original directors, Peter Rosling, a kindly, amusing man, the others involved were harder to endure. In 2003, when the whole project became the Wildwood Trust, a new and insufferable chief executive was employed and I resigned.

I sold my house in Kent and moved to a smallholding on the western ridge of a valley rolling off into distant Dartmoor, near to the tiny hamlet of Broadwoodwidger in Devon – named after the broad wood of the Wyger family – to work on a variety of reintroduction projects for water voles, teach nature conservation courses, guide tours to some amazing wildlife locations in Europe and, in the event that the forgoing could not in combination produce even a basic income, return to landscape gardening, which I had rather enjoyed. Although I of course yearned to see my dream flourish and flower, I had no means whatsoever of enabling it to do so and anything beyond the development of even a modest existence back then appeared quite unattainable.

CHAPTER THREE

Not a Lark or a Lizard Lived There

THE SNAKE-HEADED CLUSTERS OF THE GREYLAG geese look like hydras in the long grass. Orange billed and dark eyed with light surrounding rims, they turn together when sitting to watch you pass at a comfortable distance. Come too close and they will fly, exposing their broad white tails and orange feet. Although the black-necked Canadas with their white face flashings, which followed them to us, have a light tail pattern too, theirs is banded with black at its tip and top where it meets their main body. They are a darker goose overall, more ash brown in colour with black legs and bills. Occasionally an oddity appears, an escape from elsewhere. Dapper barnacles like small city clerks with bowlers, waistcoats and plumpish pale cheeks; exotic, tawny Egyptian geese, sand fawn with Cleopatra's eyes and bottle-green wing tips or a snow goose frosted crystal. No shooting occurs on our land and all know they are safe.

The greylags were the first of my reintroductions. Their parents were, and still are, birds from Norfolk, which, having perhaps collided with power lines or been wing shot but not badly damaged, cannot otherwise fly. They came one autumn as four pairs to breed in the large beaver holding pens in the farm's centre because the high fences deny foxes entry. In their first spring season, the females disappeared and their aggressively honking mates' necks extended and patrolled the edges of the dense rush beds into which they had vanished. Weeks later, they emerged with broods of tiny yellow, fluffy coated, black-beaked goslings. When these assumed their adult feathering in the late summer of the year, they eventually flew free.

That was nearly fifteen years ago and flocks of hundreds are now quite common. Some have extended out onto reservoirs nearby. In the summer they prefer the mown or grazed grass of my neighbours to the tall swards or pig-ploughed environments of my own farm. In the winter, as the grass dies back, they land in their flocks to flatten its structure irregardless. In the mornings and evenings, whether inside or out, their V-shaped skeins flying over my cottage to the west are unavoidably vocal. No one has complained much yet but when nature starts to burgeon it overspills. Together with the water voles, polecats and beavers descended from those that escaped, the geese now are unrestricted in their freedom to live and to die.

We have been visited twice in 2021 by individual white-tailed eagles from the release on the Isle of Wight. Secure as they are in their large flocks dotted and grazing in the green, the geese have had no reason yet to look upwards for danger. One day soon the eagles will find them. Their time of ease is ebbing.

———

In 2003 when I moved to Devon my hopes were low. Deflated in aspiration by what it seemed had become an eternal life game of snakes and ladders, and with my desire to create a breeding centre for endangered British wildlife, which I was convinced was essential, in the doldrums, I purchased a small L-shaped house in a tiny plot of land. The ranks of water vole pens in the small field that came with the property absorbed little time on a daily basis and, having been badly built by its former farming owner who was a bodger of the broadest sort, the smallholding offered diverting issues galore to absorb my attention. Sorting the drains that tracked water back into the dwelling, infilling huge cracks that were present in its outbuildings and removing a gigantic mound growing bright green outside the door of a large barn, which turned out to be the grave site of nearly ten thousand dead rabbits that had died in the course of a week when a former owner's breeding population of New Zealand whites contracted viral haemorrhagic disease.

While it was diverting in the summer that followed my departure from Kent to tiddle with these humdrum concerns, a decade of developing purpose near completely spirited away left an uneasy undertone. When

Paul Ramsey, my old Scottish beaver pal, called me one day to ask if I had a building which could be converted to quarantine beavers and other friends confirmed that they would like me to import some for them as well, I phoned the Bavarians and began again.

Further opportunities to do things that I truly enjoyed then followed in quick order. More offers appeared to guide groups overseas and in doing so reignite my lapsed childhood passion for investigating meadows, pools or rock clefts to unearth the creatures they contained. Bright blue slugs, it transpired, were a speciality of the Carpathians. Grey-headed rock thrushes with bright tangerine breasts were abundant in the Pyrenees. Emerald-green tree frogs in Croatian vehicle ruts. Montandon's many-spotted newt in the same environments on the Polish border with the Ukraine. At Čigoč on the Sava River I saw a thatched village of wooden houses and barns where a huge colony of white storks, which had built their great stick nests on roofs, were near utterly ignored by the peasant farmers that lived and worked beneath them.

Work went well and more commercial contracts created an income stream, which was stable rather than affluent. Before my mother died leaving a small legacy, I never contemplated agriculture as a realistic profession – the capital costs were just too high – but in 2006 when the old farmer decided to sell his farm at Upcott Grange just next to my cottage, I used her money in part to purchase 120 acres of wet dairy land. The farm had a broken-down steading with old cattle stalls and neck chains, a larger shed for young stock, which flooded every time it rained, sheep-handling facilities in a state of near ruin and an infrastructure which was otherwise pretty much entirely decrepit.

I acquired small flocks of sheep and a herd of blood-red short-horned cattle not because I really wanted to farm but rather to avoid the dissolution of the land, fences, gates and infrastructure that commonly follows when its owner rents its use to another. Casual tenants have no care, they just want a quick return for their cash spent and swiftly or slowly all degrades. I never thought at the beginning about using the land for nature. As my neighbours were all farming and blagging about the success of their undertakings, I just believed that as everyone else who was doing so was obviously – despite their raggedness – making a mint that I, too, might as well invest in proper infrastructure and in doing so develop some additional income. I installed new tracks to carry tractors and vehicles out into the fields. More renovation. Further knocking down, building, fencing and repurposing for a new future. By the time I finished, the sheep could trip-trap wherever they so wished without getting their dainty hooves damp.

Other projects came along from old contacts. I erected a few larger wild animal pens for film and photography work and in 2014 assisted with the production of a series commissioned by BBC2 to explain the lifestyles of burrowing animals.

Once again, I was creating intricate artificial worlds for the wild creatures which live all around us or, in the case of the subterranean species, quite literally under our feet. Underground is dark and moist. This is an inescapable reality. Although the producers of what became *The Burrowers* tried to capture genuine footage of life down under by sending intrepid ferrets into the dark with small infrared cameras strapped to backpack harnesses, it was all a bit of a flop. Their expensive kit was scraped off underground, misted up swiftly or, most gallingly of all for those not used to ferret exploration, was underused by their transporters who, after a short period of snuffling, were inclined to yawn and fall asleep. As high-definition, full-colour TV was required, artificial film sets were the only way forward. A large underground warren made from yellow sand, with tunnels and chambers for the rabbits. A dark woodland loam-feel with large sleeping quarters for the badgers. Tight tunnels with chambers of strange shapes based on the sort of features we had allowed captive water voles to create in trays of damp soil in their breeding pens. Tiny weeny straight ones with small nesting areas were made for the moles. Whole wooden buildings were erected to house these delicate film environments. Tunnels and runs made from more wood and wire mesh led the rabbits and badgers from their outdoor paddocks into the areas where we wished to film. As external footage of the water voles was also an essential prerequisite, a river tank setting with fern-studded bankings, lush wetland plants and a small stream was created for their use.

It's never that simple. The badgers, content to explore their film set without issue decided in the end to dig a burrow in their outside pen, live there instead and use the indoor area we had skilfully wrought to resemble a series of chambers clasped by the great curling roots of an oak as a toilet. The wild rabbits, difficult and delicate, fought whenever more than a pair of opposite sexes were introduced and were replaced in the end by lookalike domestics which were much more sedate. The male water vole was terrified of his grim, aggressive female who so dominated both him and their offspring that when they reached the age of 14 days they all wanted to leave home. The mole whose eyes were too small to show a satisfied expression trashed with its gouging forefeet every set we introduced it to with glee. Even when we thought all was simple and our control near complete, their ability to defeat our intent was formidable.

I purchased more land when I could, bought more cattle and a vast never-ending flock of sheep. The agricultural partition of Britain is stark. In the west where it's wet, it's all about livestock. In the east, where it's drier, corn and vegetables. I was part of a hearty tribe that kept sheep and cattle, grew only grass and was sure of its direction. Although farmers fall out with each other for the most trivial of reasons, they commonly stand together against all else. Outside their tribe lie enemies: vegans, conservationists, the government, other people with other points of view. As a farmer you wish the latter to support you in your 'buy British' campaigns and try to connect with them on 'Open Farm Sundays'. You want them to stay and spend money in your shepherd's huts or to camp in your fields with a view. You definitely want their tax payers' cash for your Single Farm Payment together with any other grants or subsidies you can subsume.

If they ask why the soil is dead, why the birds don't sing or if it's really necessary to kill beavers and badgers, you call them 'ignorant townies' and tell them to 'fuck off'.

Then you try to connect with them all over again.

Culturally when I came back to farming, it was like coming home. Far from the egotistical and odd world of nature conservation where big stories were talked and small deeds were done, it was a hearty culture where if you helped your neighbours, they helped you. I enjoyed identifying the best-performing cattle for my land and the satisfaction of hitting consistent grading targets for my sheep. We kept way too many. It was always too tempting to rent some lush winter grass on a dairy farm elsewhere, so you could buy those sixty broken-mouthed Welsh ewes you had seen advertised when you had surplus cash in the autumn. To take a punt on their survival.

When the cold winds blew in the worst of winters and they died despite the solid sugar block feeds we gave them, it mattered not much because they were cheap.

I liked going to market and eating fried breakfasts while pals told me that my bullocks that year – big, butter-yellow pied Simmentals out of my beef shorthorn cows – were the best I had ever bred. I liked to stand at farm gates and watch my calves running races with their buddies as the sun sank slowly blazing beneath the horizon. I liked, while catching obdurate ewes at lambing time, to smell the spring warmth in the earth in obscure field corners as nature reignited. I kept several breeds of sheep at different times. When their lambs were present in numbers, they were a pleasure to watch. Lithe brown Shetlands or Manx, tight woolled with their tiny multi-horns poking through. Tubby square Southdowns with barrel-tight skin. Black Welsh Mountains and Jacobs with their black-and-white spotted coats. All ran, jumped and played in complete accord with no thought for the differences in the shape, form or colour of each other. I did not mind the work or the rain or the cold or the accident I had when my quad bike hit a fence at full speed and I hit the tree on the other side. One of the more ancient farmers observed when I was able to hop again properly that you could tell that an individual's farming career was progressing when they began to move oddly, like the other old broken men.

Within the tribe I was secure, as long as I kept my blinkers on.

Some of the land I had purchased was in reality of little use for agriculture. Woodlands Farm, which I bought in 2010, would turn into a mire over winter, even if you kept the drains clear. Another old chap told me that no one who had ever owned the land came to any good. They had all tried dairy. *Cryptosporidium* had got into cracks in the cattle shed concrete, which every calf born inside contracted. We killed the gut bacteria in ours with a yellow drench called Halocur on their first day of life. While this treatment knocked them hard and the fresh lustrous glow of their new coats was replaced with rumpled, ruffled, starkness, you knew they would survive. Fail to treat and they would shit themselves to death in the week that followed. The owner before accepted their 50 per cent loss of young stock as normal.

We had grown to nearly 1,500 breeding ewes and over 100 beef cows. Calculating the expected production of lambs and calves from these gave you an approximate production total of individuals which, when then multiplied by an average price, afforded your expected income stream. It never bloody worked. Things always went wrong. A tractor would blow a cylinder head, a trailer floor fall out. It would be wetter or dryer than expected with a knock-on effect on price. Although I knew rationally that, even in a good year, we made no money from farming and that it required our subsidies and other income streams to stem the losses we made, the prospect of letting go when you are fully engaged and committed was horrifying. From my own experience, I knew full well that the 'all's well' message promoted by the farming unions, their pliable press and flunky politicians was shockingly dishonest. I knew when I thought at any depth on the odd occasion when I did so that my farming was wrong. Corrupt in explanation when, after many seasons of sales, the lambs or calves

Not a Lark or a Lizard Lived There

I produced barely covered their costs. I understood that the chemicals I used on them would kill all other life but I supressed this certainty with excuses that any action to kill their parasites was justified. My out all too often was because everyone else acted in a certain way, that no further significant wrong could come from my individually so doing and, in any case, that even if I changed, what difference would it make.

I knew when I paused to consider that hardly a lark or a lizard lived on my land. I am sure the flocks of meadow pipits, which were small to begin with, diminished even further in the very worst of times I imposed.

While this was ongoing, I was fighting, together with a few loyal friends, the case for the reintroduction of the beavers. Importing animals, supplying them for large penned enclosures, advocating at every level that the recovery of this life-giving species was essential. I kept breeding beavers in large pens on my farm where I saw the pools they built fill full in the spring with frogspawn and bring fish and birds to be. I reintroduced spare water voles into their watery worlds and orange-eyed pool frogs, and watched them prosper with pleasure. One year, a pal gave me a bucketful of baby grass snakes. I let these go in a beaver pond with my small son Kyle and watched them dive like dark darning needles down into its depths.

Ideas take time to form. To swirl and amalgamate. Nothing that clears quickly is ever right. Considered moments when you look back and think are critical. Deciding what to do in response requires care. Lingering

overlong before acting is, however, a serious fault as deci-
sions postponed for too long will one day for sure come
too late. I recalled in the years following my purchase
of the main farm that I had once seen a short-eared owl
in a ditch. Its glaring gold eyes had stared back at me
balefully as I looked down to note what it was before it
rose on soft wings to fly off with a vole. I watched when
a small flock of black-headed gulls came one spring to
swim on one of our ponds for a day or when a pair of
whimbrels fed in a wet field in that same year for a very
few hours. When the weather came cold and the ground
froze, I was astounded one morning to find a few lap-
wings running like chickens around our farm yard. The
cows had been outside while their streaming beds were
changed and had broken the rough soil just a little. The
birds were looking for any grubs they had exposed in
doing so, utterly unbothered at all by us. I shut the yard
that day and watched them when I had time. They were
gone the next morning and have never come back. One
year a pair of the wheatears that cross our farm annually
to reach Dartmoor spent weeks in a meadow prospecting
its long grass. I noticed and left that field for hay but
a dry spring extended into summer and the cows were
hungry. Eventually I opened the gate.

I have no idea what happened to the birds.

The saddest going was the curlews. In the time I
have been here they have come three times. Twice, in
the dawn of a morning, a single large brown bird rising
alone, calling, calling, calling its distinctive 'whauping'

lilt. Once, a pair stood clear in the midday sun of a newly shut-up hay meadow which we rented, where the grass was beginning to outgrow its winter sheep scalping. One watching while its mate fed in the warm soft soil before stooping to probe for a short time at her side. I have never seen them here again since and am sure that these were the last of the chicks born thirty years ago in a landscape that we have now denied quite utterly to them. Were they privileged as individuals to survive the dismembering of the wet meadows and moors, the deep ploughing and the drains? Or was their life one of eternal regret?

No mates. No home. No future.

My final straw came not with them. It occurred when we ran a small mammal trapping course for a conservation society in one of our woodlands with clearings scattered through. We had grazed a small herd of beef cows and calves there the previous autumn but had removed them when the rains came, for overwintering in straw-bedded sheds in our main farm yard. I wanted to show the course participants a range of common small mammals of the countryside and set seventy live catch traps to procure them. The following morning, we caught a single wood mouse. I could not believe it. We moved the traps slightly and baited again. The next day we caught two mice, one of which we had previously caught as it had to its irritation been marked with Tippex. It was a shocking result, and more disturbing still, although I looked in what long grass there was, I could not find a single fresh field vole run or feeding pile. This tiny grey-brown rodent, short sighted and bumbling, is the basis of raptorial food chains. The commonest of creatures that should have been.

No voles = no vole eaters. It's as simple as that. Their absence from my farm explained categorically why the barn owls and kestrels were no longer there.

It was enough. I was and am sick of their loss. I am appalled as I write this that these memories, which I know were only fleeting impressions drawn from moments of observation, remain so clear in my mind to this day. Though their passing was swift and I was so busy that I gave them no time, my memory is long and I think of their going now with anguish. I knew I had to change.

In 2018 I began deconstructing. The livestock in the flocks and herds we had been building for over a decade all began to go. Old pals and their calves. Sheep looking good. I could not stand the dispersal sales. I did not attend. I did not want to see them split from their friends, from their families. It hurt as the barns, sheds and fields emptied. People left too. Others, including my own son, were and are still bewildered by my actions. To carry on failing and lying would have been far easier to do.

Woodlands Farm adjoins Coombeshead Farm, which we bought in 2015, across a watercourse in a valley bottom. This unimproved lowland was always tangibly tanged by water mint when the cows crushed through. I fenced the cattle out and had a series of long pools excavated there as soon as both farms became one. A chance escape in 2013 from our breeding pens meant that beavers found these well lined with rush and willow and slid deep into their clear still waters. I found their first peeled sticks around a month after they had gone and knew that their freedom would be challenged.

The police were informed that their presence was illegal and sent an investigatory officer. When I explained that a badger, by digging from the outside into their pen, had helped them with this jail break, he laughed, went to see his chief constable and came back to tell me that it was an 'act of God' and, as such, beyond human jurisdiction. The officials sent by Natural England eventually appeared to do their own whinning on a fine spring day, when the woodland flowers released from their dark canopy by beaver coppicing were blooming around the rims of the by-then many more pools they had built. As we walked gloomily over a ridge to view their handiwork, one official caught for a moment by the spectacle said, 'This is beautiful', before stopping herself and returning to gloom. There must be subsections and clauses in their contracts that allow no joy. It's difficult to imagine how such blind unenthusiasm prevails. They disagreed with me as to what should be done, and I refused to collaborate with them. They went away telling me they would be back. That was eight years ago and, while I wait still with bated breath, the beavers have long ago moved on.

I believe now quite firmly that they have come back to heal. A natural blessing whose offspring have rippled out to the mouth of the River Tamar as far as its tidal pulse and found there wetlands where they can live in peace. Complex watercourses and wetlands now extend well down our valley and into our neighbour's lands as well who are mainly content. Ducks of many sorts fly in and out. Otters hunt fish where they could never before in ponds that are boiling with fry. Amphibians abound. Dragonflies, damsels, demoiselles and darters whirr freely and butterflies are everywhere. Lizards have returned.

In 2019, I did what I had never done: looked at my annual balance sheet and wrote off any livestock income. If the beasts made money, fine, but it would not be because we had a production target to achieve. It would at best be a bonus. We brought the Heck cattle up from fields down the road where they had lived for several years and let them out in to the forest. They enjoyed the experience. The bulls smashed small trees into the shapes of blown-out businessmen's umbrellas, gouged out with their horns the carefully constructed banks that the churchmen had made in the Middle Ages and dug pits in the pasture with their broad flat hooves in a surly, lust-ridden rage. They stood in water when it was warm, grazed on the high ridges where the wind blew the biting flies away and retreated into the cool of the treescapes when it suited their comfort to do so. When the rut time came in the autumn, the big bulls filled the forest with a low rumbling grumble. Neither bellow nor low – a sound of threat, dissatisfaction and warning. Nature giving voice.

Not a Lark or a Lizard Lived There

A few working weekends with amazing bands of fantastic helpers achieved wonders. They came and camped from all over, cooked big curries at night on oil-drum-based wood burners and dissembled lengths of internal sheep fences. Straight lines between pasture and plantation disappeared to merge what was separate into one single whole. All the crap of decades thrown in by the farmers and retained behind fences was hauled out with diggers, tractors and chains. We used the good wire to pig-proof the existing perimeter and the posts to make hibernacula for reptiles.

Our Iron Age pigs, a hybrid between wild boar and red-brick Tamworth pigs and which we had kept for photography for years, were persuaded out of the small pens they had inhabited with a large bucket full of chopped apples. As we dropped these intermittently, they wandered up the road with their grey bums twerking into their new forest home. They have enjoyed every moment of their life there since. Sucking berries from fruit-bearing trees and shrubs with their nimble delicate lips, wallowing on hot days in pools we have excavated or creating their own by launching their bodies like low-flying torpedoes slickly into soft swamp sediments. While making cosily complicated nests from dry grasses, sedges or rushes is a winter pursuit, in summer they root out a grass scrape for comfort. I am pained that I kept these clever animals in such limiting environments for so long. They, like the beavers, are the architects of life. From the tadpoles being hunted by dragonfly larvae in their wallows to the tree seedlings sprouting from the shit they excrete, life flows from their being. They are splendid beasts.

A small flock of buff mouflon are a substitute in their grazing for the herd of saiga antelope we can never keep. I hope they will provide a light nimble mowing pressure moving swiftly through the landscape. By not lingering long in any one place, their impact overall will be quite unlike the sheep whose ancestors they are. Slight, not overwhelming enough in an area of size to stunt the growth of small trees. The big males develop great manes and white flank saddles in the autumn. In a mixed-sex group, judged by their own vigour, they fight horn clashing battles for the ewes. If they do good, they will stay. If they consume tree seedlings in plenty, they will not.

The fat Exmoor ponies bay with dark beards, white noses and eye rings are an old aboriginal breed. The smooth paths that develop from their regular movements with their round flat-based feet are easier by far for small creatures to follow than those cut by large cloven hooves. In the end, when the summer months harden clay to cement, even the big beasts find it prudent to follow horse highways. Solitary bees mine the dry areas of open soil exposed by those hooves to dust themselves down in the summer. The ponies eat the heads of the tall grasses and thistles the others don't like, create short tight mown lawns, chomp casually the shrub tops they pass at random and debark tree types that are unpalatable like alder in the winter with their long rasping teeth. Their dung, poorly digested, gives different life opportunity.

Tiny tangerine fungi with linking web systems, fragile, intricate and bedecked with dew are beautiful beyond elfin. The ponies' ecological position, midway between browsers and grazers, results in an uneven structure where pastures meet woods, and this randomness, caused by their sort, all nature adores.

Though one of their ancestral forms once foraged as far as the cold coast of France, our black water buffalo, big humped and hairy, have no claim to our nationality. The cold lands never appealed much to them and they left Britain alone. My five range free in the hope that one day they will use their unique ability of combined hoof and horn power to gouge out the earth to uncover water. Their irregular wallows in wet seeps or springs can become rich pools and mires full of their green spreading dung. I feel cheated that my own mechanical largesse with diggers to create the sort of pools they enjoy has encouraged them to indulge not at all in significant exertion. They have slept and swam their first summer away. That they should be able to do what I desire with ease but have chosen not to do so is galling. It's my fault and I know it. No easy life on the next farm. Buffalo are very different to work with if you are used to dealing with cattle. Clever, more aware and amicable. They want to be friends. You need to form a relationship that is based on trust. Hit one and that's gone. Forever. A happy buffalo, however gigantic, is one which relaxed throws its tail over its back and allows you to scratch its curly pow while its broad black tongue curls up far into a nostril to lick out its snot.

CHAPTER FOUR

Why the Heck?

THE RED AND GREY CATTLE LORRIES FROM HOLLAND swept the green sides of the Devon hedge banks as they lurched through the potholes of our narrow farm lane. Hawthorn branches with their drooping cascades of tight cream flowers severed and fell like discarded bridesmaids' posies to litter their muddy wake. When they reached the yard next to the quarantine shed, the lorries turned slowly and, pneumatic brakes hissing, adjusted themselves back to the loading ramp gates. The overalled drivers climbed down from their high cabs and walked over to shake hands.

The vehicles rocked as their contents became aware of new scents: our cattle in the distance, vegetation, hay. Bellowing, grunting and banging combined with the soft lowing of the young calves. Hot breath steamed and dung splattered out through the air vents and trickled down the vehicles' aluminium sides.

'They are ok, we think, but one that was sedated has been down for a while now. It's best we get them out of the lorries and into the shed.'

The drivers dropped the big single doors at the back, which functioned as exit ramps, and then opened the internal partitions. Without pause, as the daylight beckoned, the first Heck bull to ever set foot on British soil banged its hooves down firm on the ramp in 2012 and walked slowly into the bed of barley straw in its shed. Like a western stampede, as more internal barriers lifted, the cows and calves flowed out in a torrent. All except one. Right at the back of the lorry, lying in the dark, a single cow lay, legs extended. Dead. She had been captured along with several others to create a genetically mixed herd which, in combination with others from Belgian nature reserves, would provide a viable breeding group in Britain. Born in the Dutch nature reserve of Slikken van Flakkee, a flat savannah landscape with tidal pool systems set into a clustered complex of islets to the south of the great port city of Rotterdam, she lived in a windswept environment of light, big skies, calling seabirds and seals. There her free-living herd had become so unused to people that immobilisation with drugs had been the only way to transport her to us.

A bellow from behind from one of her sisters, with the breed's U-shaped horns swinging up from her forehead and inwards at their tips, reminded us that the shed gates remained open. We shut them with haste and bolted them closed as she started from the watching herd, pawed the ground and, amidst clouds of straw flying, charged the feed barrier in an effort to kill us. We left her alone and, having drawn the lorry forward, attached a winch to the dead cow and pulled her out. As we looked down at her rigid cadaver, with its swelling belly and dulled, drying eyes, it seemed so unutterably sad that she had come from her once free life to this.

From a distance, the big bull was everything I had expected from viewing others of his sort in Europe. A dark mahogany body, cream muzzle markings and a long red stripe of bronze-apricot along the line of his back. Bulging neck and body muscles, lemon light horns with raven black tips. Although smaller by far and in shape not the same as his primaeval ancestors, as he tugged and shook mouthfuls of meadow hay, he was nevertheless an impressive beast.

———

So, what the heck is a Heck?

In 1928 Heinz Heck became the director of Hellabrunn Zoo in Munich and in 1932 his brother Lutz was made director of Berlin Zoological Gardens. They became convinced, or convinced themselves that, as the aurochs was the ancestor of all the modern domestic cattle, that their genes must be retained within them. Given this, they then theorised that by selecting and cross breeding the most primitive of breeds, that the aurochs could be resurrected.

Julius Caesar described these formidable beasts in his *Book on the Gallic Wars* as being:

Those animals which are called uri. These are a little below the elephant in size, and of the appearance, colour, and shape of a bull. Their strength and speed are extraordinary; they spare neither man nor wild beast which they have espied. These the Germans take with much pains in pits and kill them. [1]

The brothers took their inspiration from hunting descriptions of this sort and were well aware that Siegfried, the great Teutonic hero of Wagner's *Nibelungenlied* had hunted aurochs as well. If this all sounds daft, then bear in mind that the Nazis were not far from attaining power. The party's ideology for projects such as the Lebensborn 'Fount of Life' breeding programme, for people initiated by the SS with its goals of raising regiments of pure Aryan children through selective breeding, was about to begin and the contention of the Hecks that the recreation through selective breeding of both the aurochs and the tarpan – the extinct European forest horse – was possible chimed ideologically with their aspirations. As Heinz was amongst one of the first to be interned by the Nazis in Dachau as a political prisoner for his suspected membership of the Communist Party and a brief marriage to a Jewish woman, he was never a prominent figure in the regime. Lutz, on the other hand, joined the Nazi Party in its early years and became friends with Hermann Göring who was Adolf Hitler's second-in-command. The two were both keen huntsmen and Lutz gave Göring lion cubs from his zoo and many other large wild animals including Heck cattle for the private game park Hermann built on his estate at Carinhall. In 1938 Göring bestowed the title of Nature Protection Authority on Lutz.

Lutz continued his breeding experiments with this top-level support. He obtained and crossed Spanish fighting bulls with huge dark-eyed grey cows from the Hungarian Steppes; white park or highland cattle from Britain and Friesians from Holland. In the end, both brothers produced something that crudely mimicked the cave paintings. Heck cattle were go! As soon as sufficient numbers became available, they were released into game parks and, when Germany invaded Poland, into the primeval forest of Białowieża. There, Göring and Lutz, with other hunting chums, schemed to create a gigantic rewilded reserve which would, in time, occupy half of Poland. While few people would live within its bounds, it was their very real ambition that artificial aurochs, along with a herd of hybrid European crossed North American bison, would be hunted there by the gigantically fat Göring dressed in the olden hero's garb of a robin hood hat, green tights, a tabard and a spear.

It all seemed so entirely rational at the time.

In an article on his experiment long after the war was over, when the brothers were both back firmly in charge of their zoos, writing books, speaking on the radio and presenting children's TV, Heinz explained his efforts to recreate the aurochs:

Each of the selected breeds showed some char-
acteristic typical of the aurochs that I wanted
preserved. Success came incredibly quickly: by
the spring of 1932 the first good specimens of the
aurochs of modern times, one of each sex, were
born of this mixed breeding... The first aurochs
for three hundred years could be seen alive.[2]

Heinz argued there were no throwbacks in appearance
to the parent breeds from the race he had made and that
his calves were 'as alike as slices of bread from one loaf.'
This propaganda is quite simply untrue. Heck cattle are
coloured like aurochs and look very similar as long as
you put all the brick red ones, the black and whites,
and any hairy coated highland reversions that don't,
into burgers. Years ago, a chum of mine, the elegant,
silver-haired Dr Jonathan Spencer was at a conference
in Europe. Having hit the beer hard with entertaining
comrades the night before, he was snoozing through the
presentations being given the following morning when,
all of a sudden, the Dutch speaker flicked a slide up
which showed a poster of a Heck bull tossing a swastika
with its horns. Turning to a colleague who spoke the
lingo, he asked her for an explanation.

'Ahhh,' she said. 'It's a Dutch resistance poster from
World War II and it says: "Heck cattle: another Nazi sham".'

That aurochs were once found in Britain is, without
doubt, the case. Their remains are well distributed
throughout the island and one skull that was unearthed

from the depths of Burwell Fen in Cambridgeshire in Victorian times, and which now reclines in a glass case in the wrought cast-iron galleries of the Wisbech and Fenland Museum, has the remains of the flint axe which killed it embedded in its forehead.

It is unclear exactly when they became extinct but a partial skeleton found in blue clays of Porlock Bay on the North Devon coast by Nigel Hester from the National Trust in 1996 was radio carbon dated to the Bronze Age between 1738–1450 BCE and may have been from one of the last. Like Nigel himself, now grey moustached and retired, it was an old and arthritic male well past its prime.

In the nature reserve of the Oostvaardersplassen in the Netherlands, behavioural studies of the large free-ranging herd of Heck cattle which live there show that big bulls commonly move away from the breeding herds because, although large, they lack the stamina required to pursue the swift running cows. Alone or sometimes with another similar companion, they can live out a bachelor life in its reed beds for many years. On occasion, when the wind blows in from the North Sea at the end of a breeding season, bringing early snows and freezing temperatures in its wake and the herd bulls that are younger have burnt out their body fats and die, their absence can allow the solitary old bulls one last breeding fling. They fight again when they should not and in doing so can incur fatal or crippling injuries. An examination of Nigel's bull showed that it had suffered a series of injuries when alive. A fracture to its pelvis that could have caused it to limp and two broken ribs which had healed. There were signs of infection around a third rib which may have hastened its passing. Weakened and ill, did it lodge in a creek channel while foraging and drown or was it forced down in there in a last fierce fight by the horns of a more powerful rival?

By 1476 only two herds of aurochs remained in the personal possession of the Polish royal family. Eventually only a single herd grazed in the Jaktorów forest where in 1559 the Swiss naturalist Anton Schneeberger went to see them. He said that the calves were born chestnut and that the young bulls changed their coat colour at a few months old to black, with a white eel strip running down their spine. The cows on the other hand retained their hue as a reddish-brown. Both sexes had light-coloured muzzles. Schneeberger said that the 'aurochs is not afraid of humans and will not flee when a human being comes near, it will hardly avoid him when he approaches it slowly. And if someone tries to scare it by screaming or throwing something, this will not scare it in the least but while it stays in its place it will actually open its mouth, widen it and close it again quickly, as if it is making fun at the human for his attempt.'

Their end, when it came, was abject. During the reign of Sigismund III Vasa, the Polish monarch who governed the reserve, a mere clerk's record without name or ceremony in a document from 1630 stated simply that, 'In the last report [from 1620] it was written that there was one aurochs cow, but now the inhabitants of this village said that she died three years ago.'[3]

From his time onwards they were dimming. Too hard to live with. Too in the way.

———

Livestock breed societies are never keen to acknowledge any sort of image for their cherished wards which falls short of perfection. Psychosis or deleterious issues of any sort are never discussed. Like the best of crowding chanters supporting their team of whatever sort, they can be relied upon (in much smaller numbers of course and without the ability to chant – it's more of a reedy whisper) at agricultural shows to recline under banners depicting fat black, white or brown cows to proclaim their breed of choice the best. I can never remember seeing a banner at a single one of these events ever saying, 'Buy a blue bull and fit zips in your cows to get the calves out', even though everyone farming cattle knows that it's a common end result.

Many years ago, I attended an inaugural sale in Lanark Market for the British Limousin Cattle Society. Although no more closely linked to its aurochs ancestor than any other modern cattle breed, these toffee-coloured beasts from France with great wobbling butts could exhibit behaviour which was near to primeval in its level of violence. They were newly imported to Britain to increase the growth rate and weight-gain ability of our native breeds and, although their twerking abilities were impressive, it was their uneasy tempers that captured headlines in the farming press. Before the sale began, the breed society chairman stepped up into the auctioneer's box to say a few words. Smoothing his sheets carefully, he fixed his glasses firmly to his face and began to read:

'These cattle are the future. Although large, they are well proportioned, hardy, have a swift growth rate, are easy calving and have great marbling in their meat. Buy Limousin today. Oh, and PS, their temper is fine, no issue at all, just fine, buy some and find out for yourselves.'

To a ripple of light applause, he stepped down from the box, moved round to the spectator's side of the ring and the sale began. Now, for those of you unfamiliar with the layout of a cattle sale ring, they are roughly circular affairs, around thirty feet in diameter, with barriers around their circumference of around six feet (wooden if old, metal if new). On one side of the fence, a system of pens and gates allows cattle in one way and out the other, and between the entrance and exit, the auctioneer's box or rostrum is placed. The auctioneer stands and looks for bids, shouts out the rising prices loudly and bangs his hammer down firmly when a sale is confirmed. Active bidders generally stand with their elbows placed on the ring's barrier, facing towards the stock for sale, with their bums poking out behind them. Sometimes there is sawdust spread in the ring. Sometimes not. In those days, commonly old cattle traders, or 'dealers' as they are more generally known, slouched inside the cattle ring in white dust coats and green waterproof leggings, pretending disinterest while stabbing livestock with sticks.

The first heifer entered the ring like a tempest from Torremolinos. A dealer who moved his stick in self-defence was scooped up swiftly by her tabletop head and dumped, full square, on top of the startled auctioneer. As his comrades turned like rats and ran for the slim escape gaps in the ring's side, she pursued them snorting through dilated nostrils, tail up, head down with her front legs outstretched to destroy. The audience behind the ring sniggered at this spectacle and one wag shouted at the auctioneer, 'Will you guarantee she's a hand milker?'

Deprived of his microphone, which the dealer's downfall had destroyed, the auctioneer could only grin back weakly and make a V-sign. The ring was now clear apart from the heifer. 'Let her out now! Out!' the auctioneer screamed as he gestured at the yard staff with his hammer to open the exit. At this movement she stopped, panting hard and perspiring, then turned, ran and jumped. Straight at the ring barrier, straight at the crowd. She hit the old wooden top rail full bore, which cracked and fell through into the spectators. As she bellowed and tried to right herself, those who could ran. Some tried to save the lame. Some did not. On my way to the top exit, I looked back at the heifer, which having got up after flattening a few more punters, had tried to jump back through the weigh scale. In doing so, she got her head lodged and, when the knacker came to end her career with a rifle ten minutes later, trapped as she was, he did so with ease.

The crowd safe outside laughed and joked. They knew it was a big story, which they could elaborate and tell for years to come. The breed secretary joked not at all because when the heifer hit the rail and came over, she had landed right on top of him and broke his neck. He recovered in hospital but wore a neck brace for life. And his breed of choice thereafter? Well, that changed to the old rich red-roan of the native beef shorthorns, which, more placid by far, turned out to be his future after all.

———

The Heck cattle we imported were not mine. The Wildfowl and Wetlands Trust had decided to acquire them for a series of new exhibits they were constructing in the mid-2000s for their Slimbridge centre in Gloucestershire and asked me to import them from Europe for this purpose. They wanted to move away at that time from the rather dated 'ducks are us' image they had acquired in the lifetime of their founder, Sir Peter Scott (the great conservationists and wildlife artist who helped establish the then named World Wildlife Fund), to a point where they focused on explaining the importance of wetland ecology in its broadest sense. It was a far-sighted approach marred only by the reality that they had very little experience of managing any difficult creatures much bigger than a hissing swan.

Why the Heck?

At that time, I did not think they would be any more or less of an issue to maintain than the more aggressive breeds of continental cattle, which are commonly kept by farmers, such as Limousins or Blonde d'Aquitaines, both of which have a feisty feel. They went through their quarantine period without issue and spent their first winter with me while the trust planned their pens. Two of the cows from Overflakkee were just swine. Enter a shed and they came straight for you with no provocation required. No reason. No need. When we put the herd out in some paddocks nearby, they behaved no better. You could not walk near them. Without a vehicle, you dared not go through. With a deep roaring bellow, the worst one would always stop grazing, turn and race in yet another attempt to ensure your death. It was sad; she looked splendid but, in the end, sausages were her fate, though not before she so impressed one of the Wetlands Trust's directors on a viewing visit with her violence that he decided there and then to pull the project. He had an alternative. Visit Slimbridge today and you will see it. In a reed bed semi-hidden near the pen that holds the otters is a life-size iron aurochs with a small sign on its side. The Slimbridge exhibit is tame and assured. It will assuredly harm no human until one day it rusts and falls over.

I decided to keep the steadiest of the Heck cattle. They were a unique herd in a British context and it seemed, having gone to all the trouble involved in import, to get rid of them would be wasteful. Despite offering some to many who have talked a good rewilding tale, only a single individual ever took a few. His patience was not as long lasting as mine and those he acquired are now dead.

Bambi was born in the mud in June 2021. His mother, a young Heck cow, had little milk. On the first day we saw him he was following her, lowing lightly. His flanks were slightly sunken, but his eyes were bright so we left him alone. On day two he was lying on his own in a rush clump. He had little milk is his belly but his eyes were still bright and, as the lying alone behaviour was normal, we left him again. By day three he was covered in mud, it had rained heavily overnight and he had been staggering, trying to follow his increasingly wayward mum. His chances of survival were bleak, but what were we to do with a Heck bull calf?

We discussed this and then caught him. Small, fawn and leggy, he filled the footwell of the pick-up. We drove home. In the end we decided to castrate him – as a tame bull he would have no future when he became big and dangerous – and keep him for conservation grazing when he grew up, if he was placid. He has lived since that summer in a small field next to my cottage with a small dry shed and warm straw. His milk time is near over and, as he has grown, it's been a relief to find he is big, boisterous and bouncing. His small horn buds are soft and he butts in an affectionate manner.

The Hecks form a striking picture on our farm. But they are difficult. When I view them now, as I do most days in their large fields next the woodland, the old beasts form a mature mahogany mass. The great bull huge and dark. The young cows are honey yellow, flicking flies off with their tails while the small calves race and run. But I have children and visitors, and they cannot be killers. Obliged as we are to test them for TB on an annual basis, every season it gets harder as they become less compliant. They are a burden, which increasingly I do not wish to bear.

But here's the dilemma: in a world where our herds number near a billion, can it really be that one field, one valley, one forest or landscape cannot ever host a few cattle that live free?

They are beautiful and grand. It will be sad when their end comes.

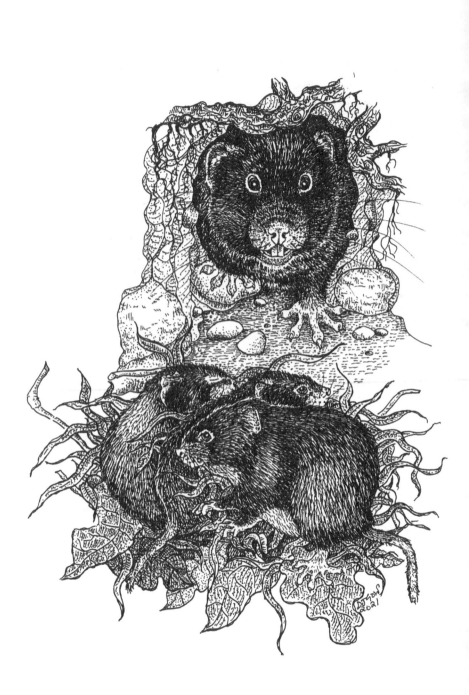

CHAPTER FIVE

Requiem for Ratty

F IVER HAD FIVE LIMBS OR, MORE ACCURATELY, FEET. Four were in the right position while the extra one jutting forward from his left front leg could also move, amazingly. It enabled him to swim like hell. Super fast. In a circle. Quite what purpose his unusual adaptation was meant to serve was unclear. Perhaps it was to confuse attacking mink or hypnotise herons. In any case, it would only ever have worked once as, although mobile on land, getting into and out of any swimming receptacle however shallow was tough. Exit was near impossible unless, through floating motionless the water itself carried him to the edge and he was able to clamber free. Fiver lived his three-year long life out in a large run on our front lawn. We never bred with him. There seemed no point. But, as he was male, his life of mowing its sugar-rich grass daily and snuggling on his own into a warm snug winter nest was lush. He had no vicious mate to run away from – male water voles commonly run along banks and push adult females into the water for safe vole sex when they are dazed – and no kids – females can have five litters a year and produce over twenty offspring – to annoy.

Water voles are bolshy aggressive rodents. Family members fight instantly with any voles from other locations which move into their living space and quite readily with weaker or less dominant members of their own colonies. While an attacker will rush in to bite any part of an opponent it can reach, defending individuals will commonly either jump and run or, if they feel bold enough, turn, stand on their hind legs and, with a hissing squeak, lurch towards their foe. While I have been bitten several times, my staff who work with them are punctured much more frequently. Once the vole has stapled its large orange incisors through your flesh it will attempt with its dark brown paws to clasp firmly the part it is attached to – normally a finger or hand – in an effort to hang on. If you open your mouth to scream, then it's best to do so with your face pointing away as the stream of projectile urine they squirt when excited will fly several feet.

Unlike most of the small creatures we share our island with, water voles attracted our attention long ago. We consumed multitudes of them in prehistoric times and threw their bones in our middens. Well into the Middle Ages, the fen dwellers ate marsh rats and Oliver Goldsmith in his *History of the Earth and Animated Nature*, reprinted as *An History of the Earth and Animated Nature* in 1855, recorded with snotty English disdain that the 'black water vole feeds on frogs, fish and insects and in some countries is eaten on fasting days.'[1] They had local names such as water rat, crabber, water dog, ratty, the British-beaver and the 'earth hound'. This legendary last description of the black form, as believed by the rural folk of Aberdeen and Banffshire, was supposed to frequent graveyards in order to devour the souls of the dead! Although their skins, like most wetland rodents, have a fine soft underfur, which they preen with great care to maintain its waterproof qualities, they were never persecuted as a fur bearer in Britain.

In the south-west of England their old name was the 'campagnol'. The old people in pubs – farmers, fishermen, land workers – recall their once ubiquitous presence still. Water voles would commonly sit on the slate shelves of the little brick butter wells built into the sides of the deep Devon lanes, where clear springs flowing through their base cooled the milk, butter and cheeses when these were still produced on farms. These features were protected from livestock with small oak doors, which, when opened, would cause the voles already inside to jump into the water but not before a glimpse was gained. This odd local name is almost certainly derived near directly from the Breton word *campagnol* for vole, which, in its own right, comes from the French word *voleur*, which means to steal, rob or filch. In former times, when withy beds were a common landscape feature designed to supply materials for baskets, furniture, fish traps or thatching, water voles stripping the soft succulent bark from their shoots would not have been tolerated. Their fondness for the bark of fruit trees, which they indulge by gnawing through the root systems until the whole tree topples, would likewise have been trying. We are never slow to notice any creature which takes anything from us and their final historic faux pas would have been to enter fields of corn, root vegetables or hops, all of which they still feed on readily to this day in Europe. Pilfering of this sort would have meant their end.

In some villages with central streams, when their numbers rose to a point where their burrows so honeycombed the banks of narrow streams that the weakened soil slid into the water and in doing so caused minor flood events or required repair, vole hunts were organised. One former participant who attended a training course I ran recently recalled these events as chaotic. The hunters – all village worthies – would assemble in the pub at lunch time on a Saturday with sticks, clubs and any available type of cur terrier they possessed, which they would tie up outside. Ignored for a time by their masters, they fought or had sex while their owners inside spent hours quaffing vast amounts of beer while boasting to amiable listeners about past triumphs. On they would waffle about spotted or Dutch rabbit-marked voles, some of which, contorted and dried with their stuffing spilt, reclined in glass cases on the walls of the bar. At a point when movement was still possible, they would evacuate in a stream-ward direction, unleashing the dogs as they went. When they got to the water, most generally fell in while some, having imbibed more than others, achieved only snoring on the sunny bankside for the rest of the day. Those still capable would lurch in a line, cursing and flailing up the watercourses, bashing banks, each other and any dogs unwise enough to come close with clubs. When a water vole appeared, uproar ensued with further injuries being incurred as a result. At the end, bloodied, beaten and bruised, the band would return with the bedraggled carcases of their sorry victims back to the pub.

Then there were children's books about the sleepy water vole or the lonely vole. Soft and round nosed, they were considered bucolic. When Kenneth Grahame wrote his book *The Wind in the Willows* in the short Edwardian dreamtime before the world sleepwalked into the horror of the First World War, so ubiquitous was his character of Ratty, so plentiful and so familiar, that the creative leap Grahame made in portrayal of the vole as a tweed-clad, pipe-smoking gentleman with an exaggerated sense of fair play was completely credible. Unlike the flamboyant toad, Ratty was reliable. His sort had always been the best of British and, with the tussle of Toad Hall won and the wicked weasels dispersed, readers were left secure in the belief of his enduring eternality.

So ingrained in our culture have they become that a few prominent individuals have their own watery tales. When the fierce Field Marshal Montgomery, the hero of El-Alamein and Operation Overlord, retired to Isington Mill in 1947 on the River Wey in Hampshire, he liked things kept neat. Tidy, shipshape and precise. Weeds were not tolerated on his lawn of one thousand square yards and he issued orders to his ex-military staff that worms and moles were to be absolutely forbidden. This obsession with order was to lead to terrible tragedy when 'ex sergeant parker who stayed on at Isington Mill as an odd job man during the last years ... was ... attacked by a horde of infuriated water-rats when clearing weed out of the mill race – and died as a result of his injuries.'[2] Monty, who was informed of this shocking event, was at the time pruning his roses. 'Most unfortunate,' was supposed to have been his only comment. Before he returned to his trimming.

In the 1950s, in a much less indulgent age, kids were encouraged – or forced – to get their butts off the sofa and to go outside to find things. The I-Spy in the Pond and Stream book awarded points for identifying different aquatic creatures. The water vole was such a common sight that seeing one merited only twenty points, which was on par with the eared pond snail and the freshwater winkle. Now you could look long and hard along most of Britain's waterways and find no trace of their existence at all. Perhaps a few old burrows here and a slumped bank there as a fast-fading legacy of a landscape once filled quite complete with busy colonies of bustling water gardeners.

———

Water voles are Britain's largest vole species. Typically, they inhabit burrow systems which they excavate with their forward-pointing teeth in the banks of rivers, ditches, ponds or streams in lush, low-lying environments. Although these structures are not always deep underground, they can, if undisturbed by either river wash or the crushing, wrenching ruin of mechanical excavators, be very old. Great-grandma houses which generation after generation of short-lived offspring tweak and adapt to their own tastes. Inside they contain many runs linked to variously sized chambers. While snuggly warm beds of soft woven grasses are preferred by rural voles and crinkly plastic shopping bags by their urban cousins, dry stocked larders of roots and hay are gathered by both in autumn to form their own well-stocked winter larders.

Over time the regular daily movements of many voles etch a highly distinctive pattern of horseshoe-shaped burrow entrances along the waterline, with round ones the diameter of ping-pong balls further up, and carved running paths and flat platforms where foodstuffs are consumed or grooming sessions occur. In the winter months they do not hibernate but rely instead on their underground food stores which they utilise during periods of poor weather. If they inhabit reed or sedge stands where no earth banks exist, they can adapt their building abilities to weaving rugby ball–shaped nests above the normal level of the water table. Mini beaver lodges in all but name.

The communal latrines created from their dark green droppings, which they deposit at the same location many times on a daily basis, can swiftly assume the appearance of mini cowpats as they are flattened with further scent depositions and constant vole movements. Water voles always leave these in visible places, such as a prominent bankside or exposed rock above the water. One of the best examples that I have ever seen was in a London canal where, on the head of a dead fox floating face down in the water, many voles in combination had collaborated to create a substantial pile of crap right between its drooping ears.

While the large vegetation, such as yellow iris or bull rush, they cut at a distinct 45-degree angle, is one of the most visible field signs of their presence, others are the creation of burrow plugs – little bundles of woven vegetation, droppings and earth – which they create to block their higher bank entrances during periods of cold or wet weather, and the distinctive star-shaped tracks left by their hind feet in soft silt. Their swimming ability allows them to both forage effectively for aquatic plant life and avoid predators by 'plopping' off the river bank to make deep underwater dives. When they pop up cork-like to swim on the surface, the vigorous paddling action of their hind feet propels them like a clockwork toy in a visibly V-shaped wake. Their leaps can be spectacular. Once, years ago, while visiting a project in a predominantly market-gardened landscape full of vintage tractors hoeing rich peat soils, where rows of clarified lettuce stretched off into the far horizon, I spotted at the roadside several water voles feeding on the top of a ditch bank. They were exposed and obvious but, as it was spring and the day warm, absorbed entirely with their food. As I stopped the truck and got out with my camera, a gust of wind blew the door shut with a bang. Instantly the voles jumped without hesitation. Up, out and down. The ditch was several metres broad and the banks were well above the water. They sailed high and soared down with a satisfyingly staggered series of splats, which were too swift to record. When I am next in the area, I will sneak up with my camera silently and thus be fully prepared when I press hard on the top of the airhorn.

Although juvenile water voles are born blind and naked, by around three days of age they develop an entirely golden coat. This changes to their normal adult pelage of a dark chestnut brown with light cinnamon bellies and cheek markings at around the age of seven days. Adult water voles in good breeding condition are approximately the size of a half-grown guinea pig with sleek shiny coats and a fat hump, which extends down from the back of their head to the middle of their back. Rotund with almost teddy-bear-like faces, they have blunt round noses, small well-furred ears, which can be tufty on occasion, and dark beady eyes.

While fifty per cent of the juveniles born in their colonies choose to remain there for life, some individuals will travel considerable distances. Recent studies indicate that this hazardous process is prompted by 'adventurous' offspring seeking not vacant habitats of high-quality alone, but rather other unrelated individuals with which to breed. As water vole populations starkly contract on a national basis, these pioneers become hopeless souls when no partners are forthcoming. They are fading fast as a British species, which, although still locally abundant in some areas like the complex ditch systems of the fenlands of Norfolk, Suffolk and Essex, are extinct or near functionally so throughout most of the wider landscape. From an estimated population of 1.2 million in 2004 – this was after a believed contraction of their British population by 97 per cent at that time – they may have declined to a recent best guess of 77,000 overwintering adults in 2021.

I have worked with them for over twenty-five years. In the beginning this was all just learning about how to keep and breed them effectively. We began to look for a source for a captive population in 1994, at a time when it was still believed they were common. I spoke to many prominent naturalists about where these might be captured and they responded with a long list of sites, which were known to contain them in the past. Knowing nothing, off we went with traps specifically built for our purpose. We found old burrows aplenty and lots of tall tales from the people we met about just how bountiful they once had been. We were always too late by months or years. Once we found a dead vole, half eaten, floating in a small cress-lined chalk stream near Rockbourne in Hampshire. Trap setting nearby produced nothing and, as a myriad of North American mink tracks in the mud under an adjacent bridge testified, it may have been the last in that colony. Eventually, a friend suggested we speak to a fish farmer near Winchester whose pond banks were so latticed with their burrows that he was glad to see some go. We set our traps on the latrines we found there and caught our first voles.

So many mistakes were made in the days when we knew nothing. They could not be kept as anything other than breeding pairs, even in the lavishly large landscaped enclosures we built full of aquatic plants, friable burrowing banks and pools. The females just attacked and killed each other. If you failed to net these structures, owls pounced at night while kestrels hovered overhead during the day. The voles dug out and escaped from simpler pens, even though we had sunk these several feet into the ground. Some individuals visibly wasted after a short time and dragged perhaps a hind limb behind them or limped as we watched. We did not understand that they had lived well beyond the duration of their natural lives and become senescent.

Again and again, we got it wrong.

Eventually we hit on a breeding cage design that worked. Six feet long by four feet high by four feet wide, this structure was built on a wooden frame with small gauge wire mesh, allowed internal space for a dry bale of straw, wood bark-covered floor and a shallow front swimming tray to allow them to keep their fur immaculate. When pairs of an even weight were introduced to each other into these in the spring, they loved their new homes and bred readily, with one amazing female in her rapid lifetime producing a record thirty-three offspring.

Fed on apples, carrots and other vegetables, they grew well and we were able to contemplate what to do next.

By the late 1990s, a series of national surveys had identified the species' precipitous decline and, in response, a broad range of research by different organisations had occurred. Conservation guidelines produced in those days focused principally on habitat creation, informed drainage ditch management, the targeted control of introduced North American mink, and education. All this effort and more has been made. We have learnt and know so much now that nothing can excuse our current combination of cumulative failure to ensure the water voles recovery.

A practical strategy is clear. Kill, systematically, the mink in the valleys where good habitat remains, landscape by landscape. Reintroduce the beavers which will bring back the complex, open, watery lifescapes they so desperately require, and reintroduce the water voles thereafter. Good-hearted organisations looking forward such as the National Trust on its Holnicote Estate in north Exmoor know this to be so and are already bringing both of these ancient partners back together. When I began, captive breeding then was a sideshow of little worth. Now it may afford the only hope that remains. In conjunction with able colleagues, I have been involved in the reproduction of nearly twenty-six thousand water voles for release projects to date. The descendants of the first release of captive-bred individuals ever attempted into the grounds of the Wildfowl and Wetlands Trust's nature reserve at Barn Elms in London still bicker and breed, nearly two decades after their great-grand ancestors were released from white plastic cages.

Elsewhere on the Bude Canal in Cornwall, at Rutland Water in the East Midlands, in the valley of the River Duchray on the Loch Lomond side, in the Cairngorms, in the catchment of Pagham Harbour near Chichester and throughout every watercourse that empties into the main River Meon in Hampshire as it wanders its way from the slopes of the South Downs into the great green swamp marsh of Titchfield Haven, water voles from our breeding programmes abound. For projects in both Scotland and Northumberland we have bred thousands of the gloriously blue-back form, once known as the 'water mole' for many years. We have reintroduced the terrestrial form into new homes in motorway embankments on the outskirts of Glasgow and know well how quickly they can bounce back if you understand them.

You need the right kind of people to help. Grim foresters, able gamekeeping organisations, sharp Belgian scientists, competent conservationists, sympathetic smallholders, children to help carry cages. Parents and grandparents to lend them a hand. People of this sort are everywhere caring and wanting to assist. Relishing the opportunity to do so time and again when they can.

It's how engaging conservation should be.

— CHAPTER SIX —

The Terror of the Tree Frog

THIS IS A TALE OF SEVERAL FROGS. IT'S COMPLICATED, but bear with me. Because it could have a fairy tale ending.

Tree frogs may once have lived in Britain. They are Europe's smallest amphibians and I keep some in a large vivarium outside my house, which I have raised from tiny froglets eating fruit flies to fully grown adults that would still find a comfortable fit in a matchbox. While one day I hope they will breed in their artificial world of pond weeds and tree stumps, they are most likely of Turkish origin and as such much less likely to prosper in our climate than the tougher tree frogs of France. Mine are like most of those kept as pets in Britain – a bright emerald green. Elsewhere in turquoise, or a more modest slate grey, they are found living free from the Channel Coast southward to North Africa, from Japan in the east to wherever their range ends in northern

Russia. During their mating season in late spring when they gather in shallow water bodies to breed, the tiny pugnacious males with their creamy brown throat sacs ascend reed stems or low scrub growth to scream through the night at their rivals. So irate do they become, like bilious back-benchers all port flushed and pompous, that you feel when they are fully wound up, they could quite easily explode.

Maybe sometimes they do.

Alone in the dark with a light popping sound.

While their piping crescendo is shrill and repetitive for most folk who live where they do, it's a normal night sound of nature. They do not have fangs or toxic poisons. Even if you lick one you will not suffer or experience the mildest of highs. Given the foregoing, their capacity to inspire terror is a novel phenomenon, which I witnessed twice in my life.

We had driven from our farm in Devon in the summer of 2017 to Martin Noble's reptile-ridden property near Holmsley in the New Forest. Martin, who was the former head keeper of the New Forest, is a great collector of both amphibians and reptiles. Low, well-cultivated pens with pruned heathers and wild herbs, basking logs and pools occupy most of his back garden. While most of his wards live there, some like the wall lizard which climbed his next-door neighbour's Virginia creeper one summer to bask in the evening sunlight where it warmed the window shelf in their daughter's bedroom – she was convinced it was an elf – do not.

The Terror of the Tree Frog

In a former life, Martin delivered Jaguar cars to customers from his family factory in Coventry before changing tack to develop a long career as a wildlife ranger for the Forestry Commission. In the end, he became the head keeper in the New Forest, which broadly involved being rude to the commoners about overgrazing issues, picking up dog shit with his hands, removing dead deer from the road and donning on every occasion of minor royal significance (along with his senior colleagues) a Tyrolean green woollen costume with red and gold crowns on the lapels. Over the years there were, it turned out, a bewildering variety of these events during which strict protocol dating back to the Ice Age dictated whether the flag flying from the Commission's redoubt at Queens House in Lyndhurst should be at half, quarter or full mast. On full mast days, to acknowledge their fealty to the crown, Martin and his mates in green mantle would march to the flag pole's foot to gaze up its flutterings from the ground. When, after a career spanning many moons, he retired, sources not close at all to the Queen generally agreed she was well satisfied with his saluting style.

Martin is a fine conservationist who, along with a handful of colleagues, rescued through captive breeding the mottled sand lizard – cream dotted brown if a female, glowing brilliant green if male – from near certain extinction in the 1980s when developers decimated their southern English heath habitats. Broad minded and intelligent, he has been a life long-time inspiration and source of great entertainment.

His wife Julia is lovely. She cares when she can for the broadest possible range of orphaned wild creatures. Foxes and roe deer, squirrels and baby birds have all been snuggly nestled by her. At one time she reared other creatures that I required for a project in the New Forest. Water deer, wild boar, wolves and red deer all became her babies.

We were on holiday. Me, Maysie, Kyle and his friend Nick. Julia was hand rearing four orphaned badger cubs in her house. They were tiny and required their bottles of replacement milk four times daily. The children were in time to help with their bedtime feed. Once they had overcome the novelty of the damp carpet soaking through to their pants – the badgers had pissed on it copiously – even the boys, farming fascists that they were, enjoyed a good badger cuddle.

While Kyle and Nick had been camping before and enjoyed it for as long as their small French beers and sausages lasted, Maysie was a novice. She was, however, keen. Very, very keen and completely prepared. A bin bag full of her multi-coloured cuddly toys occupied most of the truck's back seat, while sufficient chocolate and sweets of all sorts made the faces of Anna and Elsa on her *Frozen* rucksack look as if they were suffering from a severe case of mumps.

We extracted our tents as the sun went down, pulled out the carbon-fibre poles of the boys' blue tent, erected it with ease and pinned it into position. They moved in with beer, knives and torches. Maysie's small pink tent was next. As I extracted it from its sleeve, a swirl of shreds drifted slowly towards the ground. How odd...

Once fully extracted, the explanation became clear. Stored in the cottage loft a family of wood mice – one of which, curled and mummified in its final resting place, fell free from its folds – had spent the winter undisturbed enjoying its cosy confines. As we unfurled what was left in the vain hope that all would not be quite as bad as it seemed, it became rapidly clear that it was a tent no longer. As a mouse-inspired masterpiece in macramé, it was both impressive and unusable.

The boys in their ample abode point blank refused to allow her to share their tent on the simple basis that as a 10-year-old girl she smelt. So we went back to Martin's summerhouse, took the faded cushions off his loungers and chairs and, using these as a mattress, rolled out our sleeping bags. No sooner had we snuggled in and begun to snooze when there was a knock on the door. Kyle and Nick were weeping outside. In choking gasps, they explained they had heard a noise and, when they shone their torches outside, had seen the bright green eyes of a deer. After a quivering withdrawal to their tent, a shockingly sonic tirade had then begun nearby. Summoning all their remaining collective courage, they peered outside to see two tiny tree frogs screaming at each other from on the edge of a small green bucket.

It was all too much. They broke and ran.

They begged on their knees to come into the summerhouse and, despite Maysie telling them to bugger off, there was by then no other option. Together she and I retrieved their sleeping bags and created room in a space that was becoming increasingly cramped. Nick, however, still wept. After a period of frustrating incoherence, he eventually explained that he would need to phone his mummy so she could call him by his baby name before there could be any possibility of slumber. It was 1.30am in the morning when his sleepy and bewildered parent responded on my speaker phone.

Duckie told Pickle-pie that all would be well, that the frogs were very small and had by now probably gone to their beds, and that the deer had run away to their far forest home.

I have the recording and the photographs. One day I will produce them at his wedding.

———

The earliest of naturalists included tree frogs in their descriptions of British fauna. Described in early times as green frogs – a term which in modern times is applied to the water frog species – they were used as components in medicinal concoctions such as tooth-pulling balms and for pretty terrifying canker remedies which combined the 'iuyce of Nightshade, all the sortes of Endiue and Succorie, with Agrimonie, with Saint Iohns wort, wilde Clarie, called Oculus Christi, the flesh of Snayles boyled, Crayfishes, greene Frogges, and to conclude, with all kinde of metalls and mineralls.'[1]

The Terror of the Tree Frog

Sir Thomas Browne in the seventeenth century remarked on 'the little frog of an excellent Parrat green, that usually sits on Trees and Bushes, and is therefore called *Ranunculus viridis*, or arboreus.'[2] He was an Oxford-educated writer and physician and his Latin description literally means the little green frog of the trees. In 1658 Edward Topsell wrote *The History of Four-footed Beasts and Serpents* and illustrated what was clearly a tree frog on an acer leaf. He described his 'green frog', which had a tendency to call ahead of rain, as being distinguishable from both the common and water frogs.[3] In 1666, a Dr Christopher Merrett listed 'Rana, a Frog; *Ranunculus viridis*, the Green Tree Frog' alongside the common toad, common lizards and great crested newt as a British creature.[4]

By the time of ecologist Gilbert White in the mid-1700s, uncertainty was settling in. In a letter to the naturalist and writer Thomas Pennant, he stated that, 'Merret, I trust, is widely mistaken when he advances that the *Rana arborea* is an English reptile; it abounds in Germany and Switzerland.'

Now, people have always moved animals. The goat-like Himalayan tahr that lounge with their blonde chest beards ruffling on the high rock ridges of New Zealand's Southern Alps and the North American grey squirrels, which stick their faces into polystyrene cups of boiling soup in Princes Street gardens in Edinburgh, are amongst a myriad of species which have been transferred by people. Native initially or not, tree frogs were brought to Britain. Frank Buckland in 1845 was one such transporter, who, following a visit to the University of Giessen, brought back 'about a dozen green tree-frogs, which I had caught in the woods near the town. The Germans call them Laub-Frosch or leaf frog: they are, most difficult things to find on account of their colour so much resembling the leaves on which they live.' He put them in a bottle, which he hid in the inner pocket of his coat for the overnight train journey home. At day break, while he was sleeping, all twelve of them began calling in unison in the shaded compartment coach.

> *Well might the Germans look angry; they wanted to throw the frogs, bottle and all, out of the window, but I gave the bottle a good shaking, and made the frogs keep quiet. The frogs came safely to Oxford, and the day after their arrival, a stupid housemaid took off the top of the bottle; to see what was inside; one of the frogs croaked at that instant, and so frightened her that she dared not put the cover on again. They all got loose in the garden, where I believe the ducks ate them, for I never heard or saw them again.[5]*

Attempts to establish other colonies in more recent times were made in the Scilly Isles, at Paignton in Devon '"where none of the fifty liberated in 1937 could be found 10 years later", and on Lundy in 1933 where six years later one was still heard calling'.[6]

The Terror of the Tree Frog

In 1962 a Mr and Mrs Robinson described a visit to a pond in the New Forest near Beaulieu where they identified several tree frogs in full chorus.[7] They reported that they met several local people over the age of sixty who recalled hearing the frogs in the district as far back as they could remember. In 1964 Derek Frazer noted that this colony, which was next to a pub, was still in existence. By 1985 the herpetologist Charles Snell was lamenting their diminishing numbers in the light of no protection from collectors who wished to obtain live specimens of their own and, by 1988, although two males were noted calling from other ponds nearby, no others were noted at their long-established breeding site.

They have not been seen since.

No one knows emphatically whether this last surviving colony of tree frogs in the New Forest was a truly native relict of a once former native species or not. No bones or other physical evidence has been identified from anywhere else in Britain to suggest their former presence, so we simply cannot and may never be able to tell whether tree frogs were once native to Britain or not. We do know that other frog species such as the moor – the males of which turn aquamarine blue in their breeding season – and the agile – a yellowy dull sort which jumps very high – did exist as their remains, dating back to the ninth century, have been found. We also know that over the course of the last century, nearly 90 per cent of British wetlands have been lost to drainage and that, for all recorded time before that, every vast area which was damp had its plug pulled out by us. The great frog mires once so ubiquitous are all long gone.

We know little about the history of the small creatures that once lived alongside us because they were tiny and have left scant evidence of their being, and because for a very long time, we cared nothing much at all for wildlife unless it was big enough to hunt for fun or annoyed us by killing sheep. The naturalists who emerged as a species themselves in the Middle Ages have, in modern times, metamorphosed into conservationists, some of whom have developed thought patterns which are steeped in cultural arrogance. If there is no empirical evidence for a species having been here, then it simply can't be so, regardless of the fact that without time travel it's impossible to tell. *No evidence* means *no*. Little evidence means no and, if more evidence does emerge later, then the answer is still *no* because if you seek to re-establish it anywhere, then its population will be limited in range and thus unviable.

Or some other petty excuse.

If we truly wish to engage now, as is clearly required in a desperate last effort to ensure wildlife has a future on any basis of vibrancy, this kind of insidious thinking is fatal.

———

Though no one now debates that pool frogs – the second character in our tale – were once native to Britain (there are several bone sites, they had local folk names, were collected by the Victorians and stuffed very badly or pickled in jars) there was considerable discussion for well over a decade about what to do, if anything, as the species slid slowly towards extinction in the 1980s. They are a member of the water frog group. What this means, if you start with amusement at the idea of some frogs not being swimming-club members, is that this prettily patterned brown, black, yellow and green blotched species spends most of its time more or less on or around the edge of water bodies, while the more widespread common frog that turns up annually in most of our garden ponds gravitates towards these as spawning sites before departing for the rest of the year to more widely dispersed, soggy environments. This difference in life pattern is important because if you are a heron, egret, stork or polecat seeking slippery frog snacks, then lots of water frogs means an abundance of easy options, while no water frogs equals jolly hard work and lean pickings for your young.

Stated simply, frogs at every one of their life stages from spawn to adult hopper are, like fish and insects, the cornerstones of food pyramids for a plethora of predators. It's their natural function.

No urgency was really expressed when the last male pool frog recorded singing on its own in an isolated periglacial pool, on Thompson Common in Norfolk was collected in the mid-1990s and removed, swinging in a jam jar, by an enterprising enthusiast for cross breeding in his garden pond with some sexy Dutch females. The Wildlife Trust, whose reserve it came from, did issue a grumpy statement saying they wished to protect it – from what and for what purpose as a single remaining individual was never clear. Perhaps more pickling? Although its descendants are still present in captive collections to this day and may have been used to create other populations that are now living free, their existence for tutting officialdom is unwelcome. 'The establishment of "hybrid" populations in the wild would, at least, be potentially confusing and, at worst, could genetically contaminate and undermine the reintroduction of the northern pool frog. Therefore, the situation with captive hybrid frogs should be improved.' Which will be achieved by eliminating their existence.[8]

This official line is because, long after they became extinct, a reintroduction project starting in 2005 was begun by the Amphibian and Reptile Conservation Trust (formerly the Herpetological Conservation Trust) to reintroduce the northern pool frog to Norfolk from Sweden. Genetic studies undertaken on the pickled frogs' DNA suggested that these were the nearest remaining in type and further analysis of the distinctive call patterns of the extinct British form supported this view. Thirty adult northern pool frogs per year were flown over from Sweden between 2005 and 2008 for release at a place called 'site X' and spawn from these was thereafter collected and reared in captivity to enable the release of over one thousand tadpoles into a second site at

The Terror of the Tree Frog

Thompson Common. There, in an Ice-Age landscape well pock-marked with collapsed pingos, much work to improve the suitability of these pools was undertaken by removing overhanging tree cover to create the warm sunny environments that the species requires. That beavers could do this with ease, sustainably forever creating more and more suitable frog living space and, by doing so, enable frog gardening to cease is obvious, but unstated in the official recovery plan for the pool frog. The ambition outlined in this detailed document is to establish fifteen sites where the species exists with precisely five hundred and sixty breeding pairs in at least five counties by 2030. If a critical component of this target is that one genetic form of small frog must be replaced with another genetic form of the same small frog, then does it really matter? Certainly not to the heron that spears one through its face with its beak or the pike that swallows one whole or the otter that crunches one clutched in its soft fingered paws gleefully on a dry grassy bank. Frogs are just food for them.

Any spat about petty detail of this sort is pointless, especially when the same criteria for these riparian predators also applies to another food choice.

The introduced marsh frogs – a larger greener version of the pool frog – and their edible cousins, a hybrid between the marsh and pool frog, which having been imported in large numbers from at least the early nineteenth century as noisy, visible garden accoutrements or for Victorian vivisection, were released from Hungary into Kent's Walland

127

Marsh by persons unknown in 1935. They abound there still and have since spread rapidly in loudly 'quarking' summer masses to occupy many other wetlands in the south-east of England. Further movements have established growing populations in Somerset, Essex, Norfolk, Devon, Bristol, Cornwall and the Isle of Wight that are known and very many other breeding colonies elsewhere which are not. Hugely prolific, these turbo-charged free froggies from elsewhere are now swarming forward nationwide without targets in a leaping croaking mass. They will feed very much other life as they go and, one day for sure, will meet up with the carefully cultivated, proper little pool frogs in Norfolk which, through sexual intercourse, they will overwhelm.

Re-establishing water vole populations on a large scale has required the release of many thousands of individuals, the descendants of which now feed and function in ways that create wider habitat diversity, as they should. In addition, they provide a food resource for predators like the golden mantled marsh harriers who snatch them up with long-legged talons from their watery world to feed their nests of downy chicks. When they wake up each day, wash their faces and scurry forth from their burrows, all are potentially a big yummy supper. When beavers return more widely to Britain, as they quite clearly will, their ability to create complex pool systems of different breadths, depths and extents, filled with insects of all sorts and lined with the lushest of vegetation, will have a profound effect on amphibians. One study of the clumps of common frog spawn in a single site in Devon where beavers were introduced identified that ten clumps found on site in the first year had expanded to six hundred and fifty just three years later. Other research from Europe further demonstrates that different kinds of amphibia prefer different kinds of specific environments and that beavers, by providing a wide choice of living and breeding sites of different depths and temperatures, more open sunny wetland-surrounding habitats and a much greater abundance of aquatic, flying and terrestrial insects, are in essence the providers of ideal frog metropolises where many sorts living side by side can prosper.

Good news for them. Good news for their predators. Just, well, good news!

The only problem is that we have removed most of the other species which either were once or might have been here, and are restricting with a corset of lacklustre ambition the return of the pool frog, which is the only other we officially accept. So, as the future for amphibians, so long diminished, brightens once again it will be down to the immigrants to infill. When the marsh or edible frogs hop into habitats of this sort, they will bask in the sunshine, lapping the edges of the excavated beaver canals, splat their long tongues forth to catch the insects unwary enough to overfly them and boil the waters muddy in a turmoil of evasive movement when an egret comes along to spear them. Without these pushy foreigners, the cast returning to fill these natural larders would be threadbare and the debate surrounding whether there should be a target 560 pairs or 650 pool frogs would be the only croaking filling the air.

So, and here's the point, why not just fill the larder full to generate the best of feasts? The lost frogs like the agile and moor would be good start-up candidates for this role. Brown, green, yellow or blue to begin with, a semi-digested frog looks largely the same in a gizzard. These species were clearly here and could be restored with relative ease. We could import more northern pool frogs and cultivate them in masses in pink paddling pools for a return to new environments in digger-bucket-loads of thousands or tens of thousands or hundreds thereof if it suits better still. Why not give the job to farmers and pay them to sort it out, they after all are used to producing results at scale.

When this process is well in hand, given how cosmopolitan our own tastes now are, what really would the problem be if this great repast was given a tiny garnish of tree frog green?

Amphibian renaissance through restoration is not a threatening prospect. It's a simple suggestion with sound biological intent, and when Harvey Tweats, a self-confessed frogophile who at the age of eighteen founded his own professional amphibian and reptile breeding centre near Leek in Staffordshire, made it, he was unaware of the ire it would arouse. The established herpetological conservation community which, to be fair, is not large, slithered out to hiss loudly at his temerity. When he told one senior frogman that his vision was to restore pool frogs to every water body in Britain, the chap laughed in his face and told him that to do so would be 'disrespecting the authority of the amphibian organisation', which had spent considerable time and money re-establishing the Swedish ones in Norfolk. While I was vaguely aware from my school modern studies that a UK Atomic Energy Authority had been formed in 1954 to oversee the production of nuclear weapons, I was unaware until Harvey informed me that there was an Amphibian Authority.

We checked together on the internet and could find no mention of its existence or purpose. Should we be worried that this secret clique has already created its own site X?

CHAPTER SEVEN

Wildcats Have a Big Reputation

THE EIGHTEENTH-CENTURY WELSH NATURALIST Thomas Pennant's description of the wildcat as, 'the British Tiger; it is the fiercest, and most destructive beast we have; making a dreadful havoke among our poultry, lambs and Kids,' played to the sentiments of the time.[1] A few years ago, when my wildcat breeding facilities were much smaller than they are now, in an idle moment I presumed to just pop down to see some kittens that had newly emerged from their nests. They were lying in the sunshine on a shelf above their den. Naïve with pointy trim tails and fuzzy coats with slight stripes starting, they were not greatly bothered by my presence as I leant on the wire of their pen for a closer look. Their dad, on the other hand, perched on a nearby log pile was observing my every move. On my approach he slowly stood up and, as I got near to the kittens, he walked towards me, ears down, back crest erect, snarling softly with his ivory incisors exposed. When I leant on the wire, he exploded. In an instantly single bound he was on the shelf, which was roughly at head height, between me and his offspring, snarling and spitting, jade green eyes all aflame. Impressed by his anger, I jumped back with alacrity and thus missed the lunge he made with claws extended, forepaw against the wire. I left him watching keenly as I moved away. It was a memorable display of fury.

———

It takes real determination to annihilate small species from a landscape as large as our island. With hatred in depth. In Britain in past times, we did so repetitively. The big carnivores, such as the bear, lynx and wolf were first to go, with the smaller wildcat, polecat and pine marten diminished thereafter. We reduced the golden eagle to a much-restricted range, where it remains to this day, and, until the 1970s when its reintroduction from Norway began, the great white-tailed eagle was extinct. All of the other birds of prey, from the osprey to the slate blue merlin, were knocked back to near nothing. Otters and badgers, though now recovering well, declined significantly in the past.

This slaughter destroyed complex systems of balance where wolves and lynx predated badgers and red foxes to eliminate their competition for resources, while foxes performed the same function for pine martens.

As the fortune of one species ascended, the prospect for others declined and, on occasion, the directness of these impacts was not always straightforward. A recent Spanish study regarding the reasons for wildcats not hybridising commonly with domestic cats in Spain identified that while wildcats prefer to hug woodland/ grassland edge habitats, farm cats will opt for the rim of grassland/developed human landscapes.[2] In the landscapes between, the foxes will, if the opportunity arises, kill them both with a wary eye out behind their own being for the wolf. This uneasy mélange of influence and impact demonstrates that wildcats and domestic cats, in regions where both of their larger predators are present, are unlikely to ever find much time for intermingling in a fashion that's relaxed.

The wildcat was a widespread and familiar creature, which was persecuted in historic times for a broad range of reasons. Hunting as a pleasurable pursuit was a common medieval pastime, and dispensations for the nobility to kill cats when forest game was otherwise fiercely protected were distributed like confetti. They were not a sound quarry like the fox, buck or hare, as their habit of escaping pursuit by climbing trees rather than running was considered to be quite thoroughly unsporting.

Pursued by people, wherever they were, only once did they attempt to turn tables. A colourful tale from fifteenth-century Yorkshire records how a knight named Sir Percival Cresacre was returning home from Doncaster late one night when he was attacked by a wildcat.

The creature sprang out of the branches of a tree and landed on the back of his horse, which promptly jumped forward, threw its rider to the ground and ran away. The cat then turned on the knight and a long battle between the adversaries then followed from Ludwell Hill, where the attack occurred, to St Peters Church at Barnburgh. The conflict was so exhausting that when Sir Percival, at death's door, reached the church porch he fell onto the cat, killing it with his feet against a wall. Cat Hill nearby is believed by local people to be the point where the famous 'Cat and Man' fight started. But, while the stone of the church floor remains stained red by the blood of their battle to this day and a carved cat lounges on his feet, the real Sir Percival does not occupy that tomb. He lived well over a century later and, in all probability given their scarcity by that time, died without ever encountering a wildcat.

In 1566 the passing of the Act for the Preservation of Grain raised bounty payments for the cadavers of creatures considered injurious to the production of crops. Corn, root or livestock, it mattered not, any thieves that took them were vermin. The bounty payments due to their killers were dispensed by the church wardens within their parish. At a time well before refrigeration, when pickling, salting and smoking were the only tedious options for the winter preservation of meat, fresh rabbits were a highly desirable commodity. As a result, most medieval estates or manorial houses kept large, carefully cultivated enclosures or warrens full of these tender creatures, which had been brought north from Spain by the Normans in the twelfth century. As wildcats are better stealth hunters by far than badgers or foxes, their consumption of rabbits was intolerable. In the Devon parish of Hartland between 1629 and 1699, 311 wildcats were killed. In the parish of St Bees in Cumbria it was 142 between 1684 and 1786.

The bounties for their bodies went up as they diminished until a beast that began being worth pennies attracted shillings in payment at its end. While in 1726 Daniel Defoe noted in his journey between Naworth and the border that the landscape contained 'no inhabitants but wild-cats, of which there were many, the largest I ever saw'. And Thomas Bewick, the naturalist, corroborated in 1790 that those in the north were of a 'most enormous size' measuring 'upwards of 5 feet' from ... its nose to the end of its tail', wildcats by their time were diminishing fast.[3]

In 1843, a 'fine specimen ... killed near Loweswater was believed to be the last authentic record' from Cumbria.[4] It is quite credible that, as the species became rarer, hybridisation between those that remained and domestic cats would have occurred, and a large male, which was shot in a Lincolnshire wood in 1881, was believed to have been one of this sort. Another pair of large fierce specimens killed in Westmorland, on the Lancashire border, in 1922 were likewise considered the same.[5] No skins or stuffed specimens of wildcats collected from beneath the Scottish border are believed to exist and, although taxidermy collections in various of the National Trust's country houses and in local or county museums do include wildcats, these along with the heads of ibex, tigers or one-horned rhinos could have come from anywhere. Although there are records of wildcats being killed in Cumbria until the 1870s and others may have lingered on elsewhere, authorities of the time assembling their remains for taxidermy were stating that they had 'no faith whatever in this animal having been a true Wild Cat.'[6]

Any accurate date for their English extinction is difficult.

Though naturalists' records afford some degree of precision, the recollections of other rural observers can often afford additional recollection. In 1994, when I worked in the New Forest, Fred Courtier, a former head keeper, recalled how his father had told him of an old West Country entertainment of catching feral farm cats, which were then put in a barrel with ferrets in order to see how many they could kill before they were overwhelmed. On one occasion in around 1900, they obtained a large, fierce, tabby cat from a local gamekeeper. This cat killed all the ferrets they had and was eventually clubbed to death by the spectators. Fred's grandfather allegedly held it at shoulder height by its blunt-tipped tail, while its front paws dangled on the ground. His dad was told that it was a wildcat, and that the species was now so rare that this was the last he would ever see.

The species was forced by its killers by the beginning of the twentieth century into a mountain redoubt on the

Atlantic coast of the Scottish north west. Its salvation of sorts came from human conflict. The First World War, which saw an estimated fifty thousand gamekeepers depart for the trenches, saw the return of perhaps only five thousand in its aftermath. The wildcats, along with other targets, drew breath. The constitution by government of a Forestry Commission to create a national reserve of standing timber helped their prospects for a time. Landscapes that had been scalped bare of trees for millennia were fenced and replanted. Though the cultivated ranks of North American conifers soon grew to shade out the ground and form sterile cathedrals – 70 per cent of Scottish forest cover remains of this sort – the tall vegetation which flourished within their bounds for a short while at least afforded cover for voles and rabbits aplenty. Physically constricted by their seabound location where they could move neither west nor very much further to the north, out they moved from the mountains to the east and south, down into the lower lands.

Typically, as their populations expanded, the males pioneered, mating where there were no other possibilities with the cats they met there. The white ones and black. The tortoiseshell and grey. Cats that were ours. While many of the hybrids these pairings produced were so similar to their wild parent that the species was once believed saved, we know now that this hope was hollow. Recent assessments suggest that out of a free-living population of wildcat types numbering around four thousand in Scotland, only between 1 to 10 per cent are pure.

The anomalous situation regarding the differentiation between pure wildcats, which are protected by law, and their hybrids, which are not, means that gamekeepers can still kill them with impunity, the rabbit population upon which they used to rely has been decimated by myxomatosis or viral haemorrhagic disease, and deciduous forests, which used to be their true home, now comprise only around 17 per cent of Scottish woodland.

This situation is not good.

When I worked at Palacerigg Country Park, at the instigation of Dr Andrew Kitchener of National Museums Scotland, I set up a stud book to record the wildcats that were then in captivity. Andrew had at that time developed a system of telling wildcats from domestics by looking for specific coat characteristics, bone and internal organ features, which distinguished both domestic cats and their hybrid. This work has in recent times been supported by more detailed genetic analysis in both Scotland and Germany. So if you come across a large, broad-headed, free-living cat with a club-tipped tail whose body stripes are not amalgamated by a straight black line down its back and whose stripe pattern does not blotch or mottle run over on the A9 between Perth and Inverness, staring back at you from a cage in a zoo or in the much more unlikely scenario of an ancient pinewood up north, then it's just possible you may have encountered a wildcat.

———

Wildcats Have a Big Reputation

No standing stone, hill or memorial of any sort marks the spot where I fought my wildcat battle in a small, seedy zoo in Argyll. Its owner, Malcolm Moy, was a flamboyant character. A big man in the small world of ornamental duck keepers. When he opened a waterfowl park near Inveraray on the Scottish west coast in the late 1980s, Malcolm little realised just how tasty his wards would prove to a rather large guild of local predators. He erected an extensive fox-proof fence at huge cost only to discover that both the things that flew over it – buzzards, sparrowhawks, peregrine falcons and a bewildering array of huge gulls – together with the things that climbed – pine martens, generally, but an odd wildcat hybrid as well – and swam under it – otters and mink – remained able to help themselves to whatever dinner they desired at any time of their choosing. Needless to say, this made something of an inroad into his original collection which, by 1990, was beginning to look somewhat threadbare. With a fresh tourist season looming, Malcolm's survival strategy was to build some new cages and stuff them full of whatever unwanted creatures he could find from other zoos before, in a final stroke of genius with a tin of white paint and a black indelible marker pen, changing his wooden signs for the Argyll Waterfowl Park into the Argyll Wildlife Park in a single day.

Transformation complete, he acquired wildcats. Four kittens from another zoo. They were small when they arrived, but once fully grown Malcolm asked me if I would visit to sex them. I had no safe crush cages or traps but in those devil-may-care days, that mattered not at all. What I could do was catch them in a self-closing net. Now this nifty device operates through the use of steel wires fixed into its handle which can be pulled swiftly shut by its operator, leaving any creature captured in its soft mesh bag. I had two new ones where I worked and could pop up to see Malcom one weekend when I had time, catch the cats in their cage, pin them down in their bags by simply winding the net tight until their genitals could be seen, then transfer them into another pen and catch the next.

Small cats. Easy job. What could possibly go wrong?

As it transpired, quite a lot. When I gathered my catching gear together the day before, I included a large pair of heavy leather gauntlets, which reached up to my elbows. I was sure I would not need them but chucked them into the box all the same. We used these normally for catching foxes or badgers and, while they made our movements clumsy, they helped avoid nasty bites. I drove to Malcolm's park, met him in his cake shop, ate some large gateaux and had a happy chat. When we had done, we wandered off to the wildcat pen and began to catch cats.

Looking back, it was obvious what sex they were. Three small slim females and one gigantic, snarling male with a shoe-box square head. I should have just told him that, thanked him for the cakes and driven home. I did not. Into the cage I went. It was full of obstructions – ferns, rocks, logs on the floor – which all made cat-catching difficult, and there was something wrong with the net. Normally it opened and closed smoothly to form a tight purse. As it was, it did neither. Force was needed to shut it and even when theoretically closed, it simply was not. A large hole pretty much cat-sized afforded a means of all-too-obvious escape. While I later discovered that a colleague had used it to catch wild boar piglets moving swiftly earlier in the week and in doing so bent its mechanism, I had no idea at the time what was wrong. In any case, all went well with the females. Once they were in the deep soft net, it was easy to grab them with the gauntlets I had prudently put on, twist the net tight, sex and move them into an adjacent pen.

We left the large one until last. Alone it had sat while its mates were captured, looking down with utter disdain upon the proceedings through its slanting green eyes. I tried to net it on its perch, but it slipped away from me with a contemptuous languor and it used every piece of cover to avoid me. It was fast as hell. Though I partially caught it twice, it was simply too swift and would turn with slick ease and flow back through the opening in the net. For my last sweating effort, I was on my knees when it ran right out from under a rock and into the net. It turned, wriggled out once again, and, without thinking, I caught hold of its broad, furry tail and held on.

Bad things always happen in slow motion.

As I sat back and attempted to stand, it swung towards me and, though I just missed its intended embrace of my face, it turned and, as I lifted it higher, climbed the length of its own body up onto the gauntlet with which I was grasping its tail. Once its back legs had a good grip it looked me in the eye, snarled startlingly close and, as I attempted to grab its neck scruff with my other gloved hand, avoided that move with ease and clasped its front legs around my fist in embrace. Thus positioned, it began to bite. Vigorously. I have no idea for how long, its teeth pierced through the thick leather with ease, the leather of the glove, my skin and fingernails. They stopped at my bone when they hit it. In a dream I saw its black and cream balls tucked like small furry hazelnuts behind its back legs and, as Malcolm's face, wide eyed and pale on the other side of the wire, registered in the background, said, 'It's a male!' and let go.

It held fast to me for what seemed like eternity. Eventually in a single, swift dislocation, it turned grinned again and dropped down to the ground. I had no idea where it went. I did not care. Profoundly shaken. I sat down.

The fingers of the gauntlets were by now dripping red. They felt warm, damp and sticky. Malcolm, who had been watching from the safety of the outside, asked if I was ok? Softly I said that I was, but that if he did not mind, I would now like to go home. Cup of tea? *No thanks.* Cake? *Best not.* Did I want to wash my hands?

Nothing was going to persuade me to take off those gloves.

I drove towards home very slowly. When the throbbing all over became too excruciatingly much, I stopped and tried to remove the gauntlet. Stuck hard to my hand by clotted blood and broken nails, it would not budge. The pain of trying to remove it was much worse than not. I left it alone. The journey back to Biggar over two hours away was hell. Eventually, when I got back, my generally unsympathetic mother, on seeing my by-then grey-green pallor and blood-splattered stump, promptly rushed me to the local hospital. Tetanus jags, painkillers and blessed relief flowed. In a numb sort of way with the drugs doing their best, I watched through a kaleidoscope of changing colours the gloves' leather being soaked quite thoroughly before the nice nurse cut it up. It took a bloody long time to remove and wash, stitch and tape my fingernails and hand. It was a bugger of a mess.

———

At a recent wildcat meeting, the level of caution displayed by its participants was simply staggering. All we pretty much need to do to restore the wildcat to lower Britain is breed as many kittens from as wide a gene base as we can, do sensible precautionary veterinary screening, rear these in large pens away from people where they can climb, hunt and generally acquire life skills and stop the most murderous of troglodytes from killing them. Trial and error after that will ensure eventual success after a degree of unavoidable failure or, as it's otherwise known, learning. This obvious course of action was not really discussed. There was not time. While those from Natural England and Natural Resources Wales (NRW) said little and the chap from NatureScot's elaborate explanation was amusingly terminated by his own communications technology, most other independent attendees also voiced caution.

'We must consult domestic cat owners, the organisations that castrate feral cats, the farmers and the pheasant killers. We must explore the kind of den box designs they select in captivity, train them to hunt by using electronically powered rubber rabbits and undertake more research which, at its end, will require more research and then more after that. Whatever we do, we must definitely not breed too many just in case one day we have sufficient for release and could actually do something real.'

None of the above, it turned out, could be quickly accomplished. It would all take many years and cost a vast amount of money – approximately five million pounds has been spent to save wildcats in Scotland to date without the release so far of a single kitten – so it would be best to agree right now on a collaborative strategy of sloth.

I have captive bred wildcats for many years. It is not difficult to do. Providing they have large, open grassy pens with rock piles, logs, trees, pools and climbing branches, you can produce kittens in numbers if you have many pairs. In Bavaria, a reintroduction project that used over five hundred captive-bred wildcats to create a free-living population in the national park, which lines its border with the Czech Republic in the east, established the species successfully. Their reintroduction is possible.

Elsewhere in Europe, in the sedate mellow landscapes of Limburg, with meandering small streams lined with old pollard willows (which the beavers have found) and old, tall red-tiled farm buildings, the surplus progeny of the wildcats that inhabit the forests of the German Eifel National Park and the Belgian Ardennes have returned. After an absence of centuries, they are marching their presence westward. As the females consolidate and breed in small farm woods, the males patrolling hedgerows or tall grass corridors extend ever further. They can adapt to return to European landscapes, which are much less wooded than those of southern Britain. The return of the wildcat to England or Wales is quite possible. It's an option that's good.

The wildcat did not only inhabit forests and these environments were not only solid stands of trees. There were the cleft rock faces of sea cliffs smashed by ice and rain with sun-kissed ledges under tight crannies where their kittens could bask. Full of rock doves and jackdaws and mice, where their old friend the rabbit to this day still mows. They were tall grass fields where voles scurried under and over ground through a jungle of fine flowering herbs. They were marshes where they could lie languid on branches overhanging in the dark of a day's end, pools of warm water in a leisurely attempt to scoop unwary green frogs up and into their jaws. Great hollow trees standing ancient, spinneys of strong briar and gorse, heathlands with blowing bog cotton were once all their home.

Though a predator of small stature, wildcats have a big reputation which could easily become emblematic of life restored after ruin. To produce the many kittens required for a process of restoration that would return the species from Cornwall's tight toe to the forests of Kielder in the north will require many large secluded breeding enclosures where their inhabitants undisturbed see people not at all. Farm and estate corners would work well for this purpose in areas of woodland or meadow where space is available. A design to allow small mammals to come and to go through mesh walls and birds to fly in through their roofs would furbish their sleek feline inhabitants with a hunting experience. A choice of snug nest boxes lined with dry straw is a good thing, hollow trees and rock cavities would provide more opportunity for mothers to move their tiny kittens by mouth to different dens while their fur is still pale and their eyes are blue milk. As the sun warms and their stripes start to darken, they can greet their parents with calls and body rubs, with their buff tails ringed black held high in the long grass.

We have four pairs on our farm now in large pens of this sort. The pair we use for film and photography are well used to people and care not at all about the presence of anyone when they emerge to obtain the quail, rabbit, rats or mice that we provide for them daily. They live in an enclosure built around a tree and thorn thicket, which flowers in the spring when birds come to nest. Generally, they have two kittens and sometimes three, which, once mature, are transferred to join others in our six off-show pens in the forest above. Those that live there are more secretive by far. If you sit quiet in the evenings you can hear them 'hooting' their distinctive caterwaul.

Seeing them is harder, as it should be if we wish their babies one day to live free alongside but without us.

CHAPTER EIGHT

The Trust Went to War

T HE BIRD TOOK OFF LIKE CONCORDE, UNGAINLY AND sharp-beaked, flapping unsteadily into the blue. As I followed its course grimly from the ground, I uttered the first thing that came into my mind, 'Who the fuck forgot to clip the wings of these bloody birds?' The producer of the children's TV series we were filming shouted, 'Cut!' and turned to me to say that she really did not think we could use that piece. Could I possibly rephrase how I felt? 'Really fucking annoyed,' I replied. 'Fucking hopping! When I find out what bastard did this, I am going to ram the ruddy scissors they should have used to clip its bloody flight feathers horizontally up their arse.' Glumly, she advised that that might also not be workable.

We never recaught that bird. When I attended a common dormouse captive breeder meeting later that week at London Zoo and a fanatical birdwatching colleague's pager went off to tell him that a white stork was feeding on a salt marsh in Kent, I said nothing. I knew that Cliff and his twitching pals lived for unusual events of that sort and it seemed churlish to spoil his sensational elixir that for normal folk comes with the birth of their firstborn, a large lottery win or the inhalation of mild narcotics.

I have kept and bred white storks for many years in zoological collections in both Scotland and England. Although I was dimly aware the species had a British history – a pair had nested on St Giles' Cathedral in Edinburgh in 1416 – I knew very little about their former presence. When in 2015, as a result of a grant from the Lund Trust, I was tasked with researching what information there was with a view to advising on the feasibility of re-establishing the species as a British breeding bird, I had no idea of the tangled tale that would unfold.

Both black and white storks once occurred in Britain. They have come and gone from our landscapes for thousands of years. Giraldus Cambrensis, archdeacon of Brecknock, in his *Topographia Hibernica*, written circa 1188, recorded that, 'Swans abound in the northern part of Ireland; but storks are very rare through the island, and their colour is black.' His work included an entirely realistic coloured image of a stork swallowing an eel or snake.

Stork bones that are least 130,000 years old have been found in Devon. A bone from Roman times was obtained near Silchester in Hampshire and, in an amphitheatre of the same age near Caerleon in Wales, an excavated potsherd with an illustration of a man with a phallus-shaped nose confronting a suitably stork-like bird has been unearthed.[1] We will come back to phalli later on.

In 1507, storks appear on the price list of the Worshipful Company of Poulters at a cost of two shillings, and by the end of that century, 'White Storks were one of the most sought after of all wild birds available from the London poulterer, with prices (of between 24 and 48 old pence) comparable to those of Common Crane *Grus grus* and Great Bustard *Otis tarda*.'[2] While they were much less commonly recorded than many other species of wild bird, there are as many records of storks being consumed as there are of oystercatchers and puffins.

There are images of storks in bestiaries and in medieval manuscripts. Their carved wooden effigies project from church misericords. There are stork pubs and stork surnames. The evidence for their being is surely overwhelming.

Well, no, as it turns out. It's not. Critics of both a curt and kindly sort pointed this out as the project progressed.

The religious carvings were allegorical and, as such, referred not to the birds themselves but to their attributes of piety or care for their aged relations. The illustrations in bird books were of odd vagrants from Europe, described in their habitats by authors who had seen them there. The pubs drew their names from the fables of Aesop or were made up, as pub names are – the existence of the Unicorn pub in Coventry does not, after all, support any case for the existence of fantastical equids.

One H.D. Astley, writing to *The Field Magazine* in August 1900, noted that it was 'My custom for some years past to purchase young storks (*Ciconia alba*) from Leadenhall market in the early part of June and to give them their full liberty.'[3] Although the five individuals he released in 1900 were believed to have been shot by a local doctor, he purchased another six, 'sent over from Holland as nestlings'. These individuals, together with some older birds in aviaries, were still present on his estate on 5 October 1901.[4] A pair of storks kept as pinioned, or flightless, birds in Kew Gardens in London reared one or two young almost yearly between 1902 and 1916.[5] Storks remained at Kew Gardens until at least the Second World War when, in an interview for BBC Radio 4's series *WW2 People's War* broadcast in 2004, Nora Chivers recalled one night when her father, who worked at the gardens but was enrolled in the Home Guard, felt something sharp pressing into his back. Thinking at first that it was a German, he turned slowly to find out that it was in fact the beak of a stork.

Fair points all. Just like the case for the tree frog, the movement of storks from elsewhere to Britain for whatever purpose they served such as food, display or garden ornamentation makes it difficult to define when a wild native form was supplanted by something else.

Could it have been the case, then, that this bird was never a native and that every filament of evidence was another species, a mistake, a misidentification; or was it that we destroyed them so utterly that we have now erased near all signs of their being? Wolves were once here, too. Big, bad and bold. We built traps for them all over our islands, issued bounties, empowered nobles and charged commoners annually to make spears to do them down. When you look for actual evidence now, hard bones and teeth, near nothing remains. When we destroy wild creatures, we can remove nearly all trace. No forensics, no evidence, although we all know the felons fine well.

It's just not clear.

But there are around three hundred records of white storks in mainland Britain and Ireland since 1900. We cannot be completely certain of the exact numbers as it is difficult to determine if sightings are of the same individual bird and many small flocks. Without left rings or any other identification to explain them away as escapees from captivity, they could well have been wild. The British Birds Rarities Committee only recorded white storks from 1958 until 1983, after which time they considered the birds' occurrence to be so common that no more sightings were collected.[6] It is unlikely, therefore, that every record involves an explanation of escape and much more credible instead that when they can come blown, perhaps, off a regular flight route in Europe or inquisitively exploring the green land to the west when it sparks their high-flying attention, they do.

———

Storks have tried to breed in Britain but it's difficult. In April 1967, at least thirteen individuals were recorded on the east coast of England and two birds, residing on Halvergate Marshes in Norfolk, remained there for eight months.

They were observed performing courtship displays but no nesting attempt was ever made. In 2004 a pair of storks began to construct a nest on an electricity pylon near Netherton village in West Yorkshire.[7] This effort, which was well covered by the national press, failed when the electric company removed the nest. In 2012 an individual stork was observed building a nest on top of a restaurant roof in Mansfield and in 2014 a free-flying pair of captive bred birds from Thrigby Hall Wildlife Gardens near Great Yarmouth also failed to rear any young after constructing their nest on an old industrial chimney.[8]

For a bird occurring in low numbers when it begins to investigate a new range, recolonisation is not always easy. You need to arrive by chance at the same location at the same time as a mate of the opposite sex. You need to like each other, then build a large nest, mate and lay eggs, cope with poor weather if it comes during spring incubation, find sufficient food in a location you don't know well to rear your chick to full adult weight within a few months and then depart back to Africa before autumn.

For storks, this ensemble is further complicated because on hatching, you imprint on your birth location, so if you come out of an egg in Poland you return there, if Morocco there, if Spain there. It's like being born with preprogramed satellite navigation implanted in your brain that always brings you home and that you can never switch off. Their re-establishment where no free-living population remained in Britain was always going to be difficult. A chance example from Harewood House Bird

Gardens near Leeds showed another way might work. In the late 1980s, young white storks hatched there in captivity from flightless parents were fitted with coloured plastic leg rings before they flew from their nests. No restrictions were placed on their movements and several over time disappeared in the winter months and returned to Harewood to breed with wild birds, some of which were of Dutch origin, near to the aviaries where their parents were kept.[9] This anecdotal example of how captive-bred birds considered themselves to be British and returned to their point of birth thereafter when they had free flying choice to do so or not offered a prospect of tantalising promise.

The village of Storrington in West Sussex was recorded in the Domesday Book as 'Estorchestone'. By 1185, it was Storketon. Storgeton in 1242. In Old English Storca-tun is explained as meaning 'homestead with storks'.[10] Storrington is approximately eight miles away from the Knepp estate, whose three thousand-acre rewilding project was by that time already well established. Might its castle, large oak trees and insect-rich grasslands prove a suitable release site for storks? Charlie and Isabella Burrell, its owners, were supportive. Together with some of their farming pals, a representative from Rewilding Britain, my own staff and Dr Tim Mackrill who was working with Roy Dennis on his Wildlife Foundation, we went in May 2016 to Alsace, France, where the white stork, featured so commonly in legends and folk tales there, had become the regional symbol.

We visited two breeding centres. The first in the Orangerie Park in central Strasbourg was based in the old menagerie. Ornate barred paddocks, cramped and small, contained Cretan wild goats and other creatures collected in a random fashion. Despite this oddness, storks had bred there in captivity for nearly thirty years and the young birds, reared under a double-clutching system, removed the first eggs laid for artificial incubation while leaving their parents to naturally rear a second brood on outdoor nests, were well underway. In a dim brick room, sitting in straw-lined boxes with infrared heat lamps above them were stork squabs of various sizes. While the small ones were soft and downy like gigantic baby pigeons with long black beaks, their older siblings nearly goose size were visibly stork shaped. They moved not at all while we discussed their care but lurched, lunging on long legs as soon as a food pail containing minced meat, chicks and quail was brought in. Sharp-tipped black bills speared forward, grabbed, griped and bolted back mush before clacking sharp shut. Once fed and full, they sat and were silent.

Over time this project had reared many birds and their descendants were everywhere. On the roof of the menagerie's central brick building. In pollarded plane trees in the park. On telegraph pole tops and on the great glass roof of the Orangerie itself. Children played underneath them unconcerned on tidy lawns with ordered flower beds. Pensioners sat and snored on benches in the sun while young mums pushed their little ones along in buggies, glancing up only when their kids pointed towards the birds. Storks were wherever firm footing was found. They flew in and flew out, feeding their nestlings and tidying nests with sticks. Stopping on occasion to bill clatter to their mates before preening and returning to any tasks required.

On the outskirts of Strasbourg in the village of Kintzheim was a much larger breeding centre. Founded in 1972 its principal aim had been to return the species to the landscape in numbers and to establish a free-living migratory population. From a low point of around ten pairs when the project began, there are over 800 now. In the medieval village centre along roads lined with plane trees, many storks were present. Residents had built them wooden platforms rounded like the bottoms of vast whiskey casks with upright spikes through the bases to contain their foundation sticks. On a slab above the fire station klaxon, a hen sat calm on her nest; her mate beside her and both moved not at all when the siren blared. There were storks walking in a Cleese-like fashion through the park, storks in the street, storks in trees next to the breeding centre's excellent system of raised walkways. In the arable agricultural landscapes surrounding the village around the thistle verges and through the ploughland storks were everywhere. The breeding centre's gift shop sold storks of all sorts, on mugs, keyrings, fridge magnets and towels. They had ridiculous stork hats, the beaks of which dangled down into your face while the legs covered your ears. They look ridiculous. I bought one and love it.

Though the director of the centre expressed his willingness to assist an English reintroduction project and to provide approximately fifty to sixty chicks a year free of charge, his kind offer proved in the end hard to pin down and another source had to be sought. Warsaw Zoo in Poland is surrounded by a landscape full of storks. One quarter of the total world population arrive back in Poland in the spring. During their breeding year, many are orphaned, fall from nests, hit power lines or otherwise come to grief. Many of these are assembled for repair in the excellent, well-organised wild animal hospital located within the zoo's grounds. Some which cannot fly again can nevertheless be bred quite successfully in large grassy paddocks to produce perfectly viable young. With great grace, the zoo agreed to support our project's efforts and provide the birds we would need.

The Knepp Castle Estate led the forming project. Without Charlie and Issy Burrell, it would have been a long haul. They provided initial funding and much-needed moral support. All involved wanted to create an English Stork Project to mimic what we had witnessed in France. We wanted stork reintroduction to be not just about returning another lost species but, more importantly, to focus on the use of this charismatic bird, which will willingly live alongside us, as a catalyst for landscape and cultural change.

My study examined the birds' biology and ecology, their reproduction techniques, behaviours, migration routes and the conservation programmes initiated in response to their decline as a result of industrial farming after World War II. It transpired that there had been many successful reintroduction projects in countries such as Switzerland, Sweden, France, Belgium and Holland when, appalled by the loss of spring's harbinger, people had moved to make real their return. White storks have so long been associated with renewal and rebirth that in countries like Denmark, where their population is still very low, some householders continue to erect elaborate homemade nests on their house roofs in the hope that one day maybe, just maybe, they will attract back a bird long gone. When I discussed stork reintroductions with those involved in Europe, I asked if there had been any problems. No one said yes.

The Cotswold Wildlife Park graciously offered to import, vet check and establish two large breeding flocks of disabled birds in its grounds to supply fully fit offspring that could be released at Knepp. The Sussex Wildlife Trust's CEO Tony Whitbread offered his organisation's fund- and awareness-raising experience. Tony could see quite clearly that this could be a mould-breaking project when it came to widespread engagement. His was the idea that the Trust should seek to create living landscape-style partnerships involving rewilding at scale long before most others had considered it credible. He was confident that his organisation, which employed the stork as a focal emblem, would support a project of this sort.

Virtually all the organisations we approached for tacit comfort said yes. The RSPB promised not to sue us and even the Department for Environment, Food and Rural

Affairs (DEFRA) said – astoundingly – that they could see no significant issue. We moved on swiftly. People loved the idea. All was going well until we realised that the prospect of a reintroduction of this sort for a bird of this size on what would eventually become considerable scale was so unique a proposal that within the Wildlife Trusts ranks, on its committees and boards, disquiet was being expressed. It was a project which involved doing something exceptional and they liked the look of that prospect not at all. As discomfort rose, Tony was instructed to contact as many naturalists of stature as he could to seek their views. Tony Juniper, George Monbiot, Chris Packham and others were approached. All without exception said it was a marvellous idea. Roy Dennis wrote a splendid response:

> *Long ago in Britain, the annual migrations of big, easily identifiable birds such as white storks and cranes would have held a deep significance for our ancestors. Acting as punctuation marks in their year, they remain with us in legend, place names and carvings, even though the busy pace of modern life has largely lost that ancient rhythm.*
>
> *In this era of a monstrous loss of nature, when it appears impossible to restore on a large scale the common species of my childhood such as the yellow-hammer, grey partridge, poppy and water vole, we can succeed with some large species, especially those killed out by earlier generations for food or other*

practical purposes. It was only when I joined my daughter on field surveys in western Kalimantan that I could see how determined people could catch every large water bird for food when they desired, just like our ancestors here when they were very hungry. Removing human persecution is the key and explains why our projects on the osprey, sea eagle and red kite have been so successful, and why cranes and beavers are also on their way back.

The answers the Wildlife Trust received were not, however, the ones they sought. At worst, they expected dissension sufficient for debate and delay. At best, clear enough support for a no vote to allow them to wriggle free.

Tony was by then coordinating a funding application to the People's Postcode Lottery, which would have afforded the project an ample start-up resource. As he tried in vain to placate dissension, the stew of rebellion within his own ranks began to bubble and boil.

I was invited to address the Trust's conservation committee for a twenty-minute session to explain the case. I live on the Cornish border, four hours and two minutes away from the Trust's HQ in Sussex. When I said that I had no intention of travelling that far to explain a complex subject without having adequate time to do so, they grudgingly agreed to give me more time. Charlie Burrell kindly consented to accompany me and, when the appointed evening arrived, we assembled in the Trust's hot, stuffy meeting room. There was a good turnout and my presentation began. After fifteen minutes, Charlie poked me to whisper that he reckoned that only 20 per cent of those in attendance were listening. One had gone to sleep at the back while the bulk of the others were whispering and passing a tiny spring-tailed insect around in a test tube and gulping when it twitched.

Why bother turning up, I wondered, if you did not wish to listen?

When I finished speaking, the answer was clear. The old chap at the back snorted and woke. He had questions. Having listened to nothing, his concerns were significant. He was an ex-employee, I think, of the old Nature Conservancy Council and not a happy man.

What would happen if birds died?

Were they ever a British breeding bird?

It would not work!

If the Trust told us to stop at any point, what would we do?

Should we be doing this at all?

It's an introduction not a reintroduction?

It will spoil our bird records!

It will spoil my own personal 'whoopee' moment when I find a wild one!

One luminary went further, producing a paper called 'Clutching at Storks' with an image of some grey herons on the front. After discounting generally near everything to do with any possibility of the white stork ever having been a British bird at all and suggesting that the Storrington name derived from either herons, egrets or cranes, he additionally identified that the word 'storch' recorded in German literature as stork has a secondary meaning as penis. Could Storrington have been associated with a resident's sizeable member? To have been remembered for a willy of this sort for nigh on one thousand years, it must have been a whopper. A true spectacle to behold.

It was all too late. 'Storkgate' had started and would lead to an internal Trust war of an uncivil nature. The sharing of staplers stopped. Personalised mugs were secured and pencils sharpened. Elastic bands with projectile potential disappeared with promptitude from the post and were hoarded in stockpiles. The Dream Fund bid collapsed even though the project had been short-listed to the last few applicants, and the partnership that involved the Sussex Wildlife Trust dissolved. They called me a liar, and not long after Tony left his job.

The Knepp Castle Estate overtook all works and in the end brought in the first birds from Poland without the Trust's collaboration. Some hundreds of others have been imported since then to reinforce other free-living flocks elsewhere.

With more knowledge gained from the many other successful white stork reintroduction projects that have occurred in Europe and, backed by the will of Charlie and Issy to see it through, the Knepp project is different to previous reintroduction attempts. All the early indications are that it will work. The White Stork Project reformed and obtained sufficient numbers of birds from Warsaw to make a start. A breeding flock drawn from these was established at the Cotswold Wildlife Park while another group was fenced into a large wetland pen at Knepp. Other partner organisations with different, complimentary skills offered their assistance. In 2020, I was on site with Charlie when the first of their young birds hatched high in an oak tree in May and took to its wings for the first time. The hundreds of people watching laughed and cried. Many tens of thousands more who contacted the estate thereafter in the dark days of the plague year expressed their hope, wonder and deep gratitude for their effort to restore this astounding emblem of hope.

And that's what it is and was all about. Visible hope, renewal and resurrection at a time when natures prospects are bleak.

Charlie gave me a number of Polish storks for a colony I have now formed on my farm. Comprised in part of his birds plus others that I have acquired from zoos, they occupy a series of large sunny aviaries with tall platforms, nests formed over years, grassland and pools. Ours are calm and well settled. They too, like the Knepp birds, have begun to breed. Three chicks in 2020 and three in 2021 that lived. A white stork in excellent condition is white only superficially. At a distance. Given good diet, care and shelter, their plumage acquires a depth of roseate that can be observed only at close quarters when they move or turn to expose their underlayer of feathers. The weather is an issue, with warm springs then leading to wet May weeks at hatching. We will put covers over their nests in the autumn so that even if cold they will be dry next year. Their mothers are learning to brood them well. Sitting tight to cover their pterodactyl-like offspring when the west wind blows. Tufted and wheezing they grow swiftly in weeks, stuffed full by their parents on a diet of minced rabbits, rats, sprats and chicks until they are the weight of a turkey. In the next few years, we will have more than sufficient to release into our rewilded farm where small mammals, invertebrates and other insects will provide a full cornucopia of food.

I hope they will nest soon on the roofs of our own sheds and houses.

Perhaps one day they will meet and join the Knepp birds soaring in squadrons along the south coast to the Scillies, with Africa on their minds.

CHAPTER NINE

Let's Give This a Pulse

W E STOOD ON THE BRIDGE AND LOOKED DOWN. Kathleen, frail but upright with one hand on her walking frame while the other clasped her daughter Sarah's arm firmly, asked, 'Do you believe me?' in a soft northern lilt.

'Yes,' I said. 'I do.'

The location we overlooked on a stream beneath the National Trust's property at Wentworth Woodhouse in Yorkshire was ideal for beavers. A shallow water body with a raised stone weir flowing to the right of two small cottages that she remembered from her childhood. The stone bridge we were standing on had not changed at all. The trees in the landscape surrounding – ash, willow, sycamore and oak – were all species which beavers could and would have consumed had they chosen to do so in the early 1930s when she saw them. She was a little girl then, whose father, James Nixon, as a winding engine man in a coal mine, was responsible and intelligent. Although he did not travel, he was well read and he cultivated his interest in the natural world in his children. Kathleen remembers well the first time she stood on the bridge, looked down and saw the beavers building their distinctive dams. She can remember them carrying sticks and is sure she saw them on several occasions.

After a time, they disappeared and, despite often looking, she never saw them again. At 91 years of age, she wrote to tell me of her memories and, in early June 2021, I met her on the bridge.

Although some of the trees have been coppiced at around the time she recalled, there is no evidence of beavers and it's not likely, given the time span that's elapsed, that there ever will be. But she is a clear-thinking individual who, as a young woman with a family in the 1960s, was a keen narrow boater. She saw coypu when they lived free on the broads and knew those rodents well with their white, grizzled beards. She remembered beavers did not look like that and coypu do not build dams. There was once a menagerie with a bear pit and llamas at Wentworth Woodhouse and it's possible that her beavers may have been escapees from there.

But Kathleen's stream is a tributary of the Humber where faraway Beverley was a beaver meadow in Saxon times. The old man who told the chronicler Edgar Bogg in 1904 that his grandfather had witnessed these curious animals in 1750 or thereabouts at a site called Oak Beck near to a location called Beaverdyke and the last bounty record from Bolton Percy in 1789 are in a planetary timescale not far removed from Kathleen's memory.

While records lost or not yet found may one day fill in, it's not fanciful to believe her account at all.

Her wonder at them, recalled at the age of 91, was genuine and real.

While she could well be the last person in Britain to recall the loss of the beavers, for the rest of us less lucky, their passing is conceptual. It was a joy when she came to see my beavers in Devon in August 2021 and was able to meet at close quarters a large gentle male lying at rest in his shredded wood bed. Standing next to him, Kathleen's presence perturbed him not at all. He appeared in her eyes, to be near identical to those she remembered from long ago, complete in shape and form.

———

I once thought that the conservation of wildlife on this wealthy, ordered island would be a competent and organised endeavour. I thought the organisations, both voluntary and government based, tasked with the conservation of our natural world would work hard for its salvation. I know now that this is not always so. Although I did not watch the Tokyo Olympics, I am confident that we the British must have won bronze, silver and gold once again in the can-kicking contest. More small, grey non-entities standing together on a dais after completing, no doubt, their own tedious analysis of the complications pertinent to its ascent before stepping cautiously upwards blinking into the spotlight of interplanetary incomprehension.

Some fine individuals have tried hard to make progress, to improve nature's health, to advance and repair
in the harshest of times, but the government bodies
responsible for wildlife recovery, broken in the bouffant
years of Margaret Thatcher's government for disputing
the environmental efficacy of her more malevolent countryside policies, have ever since been maintained on the
basis of near-perpetual decline. Long gone is their bite
and any bark they once had of influence or ability has
declined to a pip squeak. Hesitant and confused, they
blow limply with every political gust, seeking to control
near complete every function but seldom to lead or assist.

An image of helpfulness is, however, good and here's
an example. As an understanding develops of the process
required to restore glow-worms – an innocuous beetle
delight – the National Trust, wishing to return them to
one of their sites, dropped a line to Natural England for
their thoughts and received this in response.

Has the receptor site ever held glow worms? *Who
knows? They were once everywhere.*

Is the site managed, and managed appropriately? *Yes,
for nice gardens and statues. Who knows what is appropriate for this species, as only by trying to restore it will we
actually find out?*

Is the site geology appropriate? *Hmmm not sure. Why
and how is this relevant?*

Does the site hold populations of nationally scarce or
rare molluscs, given that the proposal involves the introduction of a molluscan parasitoid? *Probably not known*

but if so, they are adapted to glow-worm predation and the number of glow-worms will be dictated by their abundance in any case.

The replier advised that introductions of this sort ought to acknowledge the International Union for Conservation of Nature introduction guidelines and adhere to as much of their identified criteria as was possible given that this was a local initiative. Donor glow-worms from other populations should not be mixed as this would hamper a disease risk assessment process.

The respondent went on to say he was trying to be helpful.

Well, even if you accept, with your hard hat, head torch and ropes on, that a thorough exploration of the Earth's crust will be necessary before you released a single twinkler and that their existence poses a disease risk likely to jeopardise the existence of humankind, it's not exactly an overly helpful or even particularly clear response.

If you wish to bludgeon badgers or beavers or remove peregrine falcons and hen harrier chicks from their nests, a way can be found. If you wish, on the other hand, to restore fading species for nature conservation purposes, then you have to fill in 90-page documents which will be thoroughly scrutinised eventually and returned to you with a further suite of impossibly complex questions. Why is it, when nature's ebb is so low and when so many good people want to do good, that there is no easy 'one stop' shop that tells you what is best to do with your land, or your money, or your time to help. Why can't you just cultivate endangered plants and grow then in your garden, buy packets of glow-worms and distribute them with glee, release pool frogs in your pond or adders in your meadows.

Why does everything have to be so slow?

Why can't systems be developed to move things fast?

It's how every other industry works. Farmers do it well.

As the cloud of COVID lifts just a little, I have attended a number of conferences in recent months. At one, an ebullient lady asked for funding for the Farming and Wildlife Advisory Group to map the remaining biodiversity on farmed lands. I am unenthusiastic about more public resources being spent on ventures of this sort as I believe it's simply more cocaine for an addicted industry. The time for cold turkey is now and, if you really need to undertake this sort of study at all, then trained teams of lower primates with crayons – red for the flat lands which are dead, blue for the small slopes and odd pockets which may not be – could accomplish this task with ease for bananas bunged through their bars.

Instead, the money should be spent on recovery. What do you do with a single wet meadow of an acre or less where yellow iris remain in whole valleys gone green? A slight bank on a slope where the harebells blow in the late summer breeze with wheat all otherwise surrounding. Orchids in a muddy rut next to diesel cans discarded in a ditch or tadpoles teaming in a quarry filled in with farm waste.

Joining these dots won't be easy but there are already techniques which could be refined, updated and improved at scale to create effective solutions where these are required. While the incoming political mantra of 'public money for public goods' has made clear its course to restore life-filled landscapes, and exhibited so far a steadfast intent to do so despite the lobbying of those who have had trailer loads near literally of your money to simply support their existence, resources of different sorts will be required to deliver this dream. Perhaps these will come from carbon sequestration or be wrung from developers' purses – from wherever the main flow originates, any public contribution will diminish in dilution and must therefore be well invested if a difference is to be made.

Repairing a few stone walls here or there or planting an odd tree is a charade and others with different ideals must be empowered to chart this new course. Projects like that on the Knepp Castle Estate, which choose to expand out into a grand complexity of variable environments, have already delivered huge value for tax payers' pounds. The 'New Nature' habitats developing there at random have become free willed. Resurgent purple emperor butterflies where none before were known, nightingales, turtle doves, lesser-spotted woodpeckers, nightjars and skylarks. All the British owls and thirteen out of the UK's seventeen species of breeding bats. Occasional surprises turning up such as Montagu's harriers, prospecting black storks and visiting golden oriels, which one day might just stop, settle and breed. More projects of this sort are required. Great eco dynamos that churn out more life, year on year, as the development of their lifescapes afford more food, cover and opportunity.

No more decline and delay. Instead a course of pushing, pounding, production.

While other farms of a similar size similar to Knepp, former grouse moors, the land holdings of water companies, the Defence Ministry and the great feudal estates could all assist with this sort of reshaping at scale to create nature space, it is entirely desirable that smaller land parcels should be reprofiled as well. Scene setting through scrape excavation, the reinstallation of rock piles, clever tree planting, the leaving of dead timber where it falls, rewetting, reseeding, replanting if appropriate with pollinators and structural plants are already standard conservation tools. There is no good reason why these innovations, appropriate right down to a garden or roadside scale, cannot be employed to create collaborative corridors for life.

On 17 May 2021 the government's minister for Environment, Food and Rural Affairs, George Eustice, announced that 7,000 hectares of woodland would be planted per year until May 2024 to set on course a trebling of reforestation rates. This is a big political decision and it would be good to think that his protection regime for the new saplings trembling tender in their tree guards or behind their long, tall fences has been thought through to a time beyond enclosure. As these woods start to produce offspring of their own in a cycle which should allow great trees to rise and expand over centuries, to hollow and hole, to stand the ravages of fire and the wrench of wind and rain to become overlords of a realm eternal, will they still require our tinkering and detailed protection? As it stands the answer is yes, as the deer population that inhabits our islands is nearing two million strong and expanding annually by 30 per cent. They will bloom to fill future forests full and there will be no mystery in what happens next. It is known.

We are the fools that cannot see.

One day inevitably, if we survive long enough and are in any sense sincere, we will have to come to terms with the wolf. We will have to forget our compassion for sheep alone, reinforced, as it has been, by tall tales, Christianity and seams of woollen commerce once so significant that are now long spent, and move on. Think of the great American ecologist, Aldo Leopold, who killed the wolves in plenty in his youth in the end came to realise their role full well. He understood that 'just as a deer herd lives in mortal fear of its wolves, so does a mountain live in mortal fear of its deer. And perhaps with better cause, for while a buck pulled down by wolves can be replaced in two or three years, a range pulled down by too many deer may fail of replacement in as many decades ... Hence we have dustbowls, and rivers washing the future into the sea.'[1]

Caught wild in Europe, with a well-founded fear of us, any new pioneers will have to be crated and brought back. Will they find the old dens their ancestors were dragged from in the killing times of the past as the new pairs in Holland have done? Perhaps in the scree slips of Snowdon or in the limestone karst caves of the northern dales. Maybe they will pause to consider and lope long for Rannoch where their last lair lies open in a gravel ridge high on the flanks of the Findhorn River. Wherever they lodge, they will teach deer to run. Not to hesitate, but to flee without thought wherever they can to avoid a crushing, tearing, wrenching end.

To be fearfully uncertain of their lives.

———

While the wolf is a dream, our rewilding at Coombeshead is becoming reality as the last of the internal fences come down. Within the bounds of our ring fence, the sheep have all gone and will never be back. We will reseed many wildflowers in polytunnels where they used to winter before their spring glut of lambs. Where they have not been, small oaks are now growing from a single season's rest. A pair of brick-red Tamworth pigs will root in their place to open soil and create a bed for plant seedlings and other small trees. In the wet valley bottoms and woods reopened to grazing, the wild plants once stifled, will be encouraged to emerge out into a landscape with more opportunity. No constant mowing, more random and ragged. We are cultivating long strips to restore yellow rattle to suppress the rich grasses that still dominate. In time, I hope it will reduce their roots to dust. The beef cows we have left will ponderously pulse in and out of our wild areas in mimic migrations of plodding dun horses, red and back. They are easy and will disturb only insects with their swishing tails in summer while carrying birds on their backs.

Five Konik horses with zebra-stripped legs came last week and stand, tails whisking, in the warmth of the late summer sun on top of the banks of the new pools we have excavated. The water piped for so long down deep in the earth has been brought up by the same mechanical excavators that entombed it to, once again, flow free. We have dumped large rock piles in pastures de-stoned for millennia. The robins and meadow pipits, dunnocks and wagtails are exploring them all now. Next year, when the vegetation has changed to allow voles to live in fields again, as they have not done for decades, we will reintroduce adders, grass snakes, common lizards and slow worms – I do not think the few that linger in the hedge banks are strong enough in numbers now to accomplish their own recovery. In years to come, they will wriggle in for winter rest and bask on sun-warmed stones in the summer. Wheatears may bob on their high rocks, white arses blinking next summer when they return from Africa. Maybe they won't go to Dartmoor, and stay and breed instead. We have cut now into level land and wet scrapes are filling with shallow water over winter. The only waders left now are seasonal snipe. Will any of the others long gone ever return to find them suitable? The piping redshank, wing batting peewits and long-calling curlew?

Grey partridge pairs are coming tomorrow. We will incubate their eggs next spring.

We will breed many wildcats for return to the south and more water voles and white storks in time.

We will kindle fairy lights of hope by returning the glow-worms, with the help of children, to our rough banks and verges so that those living now can walk in the dark and delight.

I want to a provide fulcrum for a movement that can supply seed. Seeds of hope, seeds of thought, seeds of training and seeds of ability. To build a network of people of all sorts and ages who will help each other when one day they meet.

Supporting the very brightest of young people in their commitment while they retain visions of their own is essential. We must not eternally seek to break them to our thinking without considering theirs at all. When we strive to mould their minds to suit ourselves, we ensure their value will be mundane – more of our same vision failure.

Once the ground is ready, then all can be grown. No magic beans. It will take time but, for people who care with ability and passion, I want to help afford the tools they require so that those who say no are confronted by more who say yes. In time, it will be our way and yours.

Let's give it a pulse of life and of hope. Let's do what we can.

Joining the Dots

W HEN THE KIDS WERE LITTLE AND THE WEEKEND
came along, we would travel to a swimming
pool in Bude. It had a good flume, which we
slid endlessly down on Sunday afternoons pretending
to be otters, seals or sharks. As we drove there, week
in, week out, we passed a small overgrown meadow
at a place called Pancrasweek. It was a tiny two acres,
but within it ferns and wild flowers grew amongst grey
willows. In June, its carpet of flowering flag iris turned
all solid gold and hid for a time other wonders. It was a
near final example of the once widespread 'river hams',
perhaps Bronze Age areas of seasonal grazing where the
land met the water and in doing so developed intricacy.
An ancient richness of interweaving plants, birds and
animals, all destroyed within memory by us. One day
there was a sign in its centre. Newly erected. Its headline
said 'For Sale' and in the small print a further sentence
suggested that it would benefit from 'agricultural
improvement'. For 'improvement' read a wide-tracked

excavator, trenches, coiled blue plastic drainage pipe, willow wrenched out, medieval banks flattened, ploughing and resowing with rye grass.

I was bloody furious!

When I got back home, I wrote a bitter tweet to which a chap called Rob Thomas replied. Rob, it turned out, had organised a crowd-funding project for an area of ancient woodland in Wales, which had been threatened by destruction and suggested I emulate his example. I was sceptical of success but he set up the mechanism and together we started an appeal. I was working in the Bavarian National Forest Park in the week that followed and, as a result my phone reception was not great, but day on day more people responded. Some offered to help coordinate. Some to warden the site if we obtained it. Children sent money in small sums. Others of all sorts assisted as well. It was so moving. It still brings a tear. I committed some of my savings and other charitable institutions insisted they would help to secure it. Jon Whitfield, a folk-singing barrister who writes poems about hares, then assisted us to secure.

We bought the meadow and it is saved.

We have formed a small Trust to care for it now and will use its seed as a source to create more of its kind. Through green haying or direct cultivation in thousands of pots. We will do more. To re-sculpt the farm and develop its potential in the ways I describe. To buy more land around us and allow all to grow. We will begin a journey, which, once started, is unlikely to ever end.

We will open to show you each year.

Epilogue

If you wish further details or wish to help us with our journey, contact Keep It Wild Trust at info@keepitwild .org.uk.

— ACKNOWLEDGEMENTS —

Nick Wainwright for being a good sport, but still showing the video at his wedding. Martin and Julia Noble for being so splendidly kind, hospitable and supportive. Sarah Bridger, Nigel Hester, Rodger Heap, Iain Valentine, George Reid, Peter Cooper, Harvey Tweats, Laura Wade, Simon Hicks, Dr Brian Thomson and Dr Carl Jones for being such good sports, and Tara Kinsey for putting up with my histrionics while writing this book.

— NOTES —

Chapter 4: Why the Heck?
1. C.J. Caesar, *De Bello Gallico* (Project Gutenberg; Everyman's Library version, 1915 edition, translated by W.A. MacDevitt).
2. Heinz Heck, 'The Breeding-Back of the Aurochs', *Oryx* 1, no. 3 (September 3, 1951): 117–22, doi:10.1017/S0030605300035286.
3. Cis van Vuure, *Retracing the Aurochs: History, Morphology and Ecology of an Extinct Wild Ox* (Sofia: Pensoft Publishers, 2005), 71.

Chapter 5: Requiem for Ratty
1. Oliver Goldsmith, *A History of the Earth and Animated Nature* (Glasgow and London: Blackie & Son, 1855).
2. Alistair Horne and David Montgomery, *The Lonely Leader: Monty, 1944–1945* (Australia: Pan Macmillan, 2009), xxviii.

Chapter 6: The Terror of the Tree Frog
1. Timothie Bright, *A Treatise, Wherein is Declared the Sufficiencie of English Medicines, for Cure of all Diseases, Cured with Medicine* (London: Henrie Middleton for Thomas Man, 1580), 44–45.
2. Thomas Browne, *Pseudodoxia Epidemica* (London: Printed for Edward Dod, 1646).

3. Edward Topsell, *The History of Four-footed Beasts and Serpents: Volume 2* (London: G. Sawbridge, 1658).

4. Christopher Merrett, *Pinax Rerum Naturalium Britannicarum* (London: T. Warren, 1666).

5. Francis Trevelyan Buckland, *Curiosities of Natural History: Second Series* (London: Richard Bentley and Son, 1883).

6. Richard Sidney Richmond Fitter, *The Ark in Our Midst* (London: Collins, 1959).

7. W.H. & M. Robinson, 'Letter to Frazer' (J.F.D. British Herpetological Society Archives, 1962).

8. J.M.R. Baker & J. Foster, *Pool Frog Reintroduction Plan for Thompson Common, Norfolk. Version: Version 20 March 2015* (Unpublished report. Amphibian and Reptile Conservation, Bournemouth, 2015): 33, http://www.breakingnewground .org.uk/assets/Projects/A3/Pool-Frog-Reintroduction -Plan-Thompson-Common-2015.pdf.

Chapter 7: Wildcats Have a Big Reputation

1. Thomas Pennant, *British Zoology, Vol. 1: Class I. Quadrupeds. II. Birds* (London: William Eyres, 1776).

2. A. Rodríguez et al., 'Spatial Segregation Between Red Foxes (*Vulpes vulpes*), European Wildcats (*Felis silvestris*) and Domestic Cats (*Felis catus*) in Pastures in a Livestock Area of Northern Spain', *Diversity* 12, no.7 (2020): 268, doi:10.3990/d12070268.

3. Daniel Defoe, *A Tour Through England and Wales: Divided into Circuits or Journies* (London: J.M. Dent, 1928); Thomas Bewick, *A General History of Quadrupeds* (Newcastle, S. Hodgson, R. Beilby & T. Bewick, upon Tyne, 1790).

4. Richard Lydekker, *A Handbook to the Carnivores: Part 1. Cats, Civets, and Mungooses* (London: W.H. Allen and Co., 1895).

5. H.L. Edlin, *The Changing Wild Life of Britain* (London: B.T. Batsford Ltd., 1952).

6. Rev. H.A. Macpherson, *A Vertebrate Faunas of Lakeland* (Edinburgh: David Douglas, 1892).

Chapter 8: The Trust Went to War

1. Iain M. Ferris, *Roman Britain Through Its Objects* (Stroud: Amberley Publishing, 2012).

2. W.R.P. Bourne, 'Fred Stubbs, Egrets, Brewes and Climatic Change', *British Birds* 96, no. 7 (July 2006): 332–39.

3. H.D. Astley, 'Storks in England', *The Field*, 1900.

4. H.D. Astley, 'White Storks Remaining in England', *The Field*, 1901.

5. Richard Sidney Richmond Fitter, *The Ark in Our Midst* (London: Collins, 1959).

6. K.A. Naylor, *A Reference Manual of Rare Birds in Great Britain and Ireland* (Nottingham: Naylor, 1996).

7. Mark Cocker and Richard Mabey, *Birds Britannica* (London: Chatto & Windus, 2005).

8. George Monbiot, *Feral: Rewilding the Land, Sea and Human Life* (London: Penguin, 2014).

9. J. Waller, personal communication, 2015.

10. Eilert Ekwall, *The Concise Oxford Dictionary of English Place Names*, 4th Edition (Oxford: Oxford University Press, 1960).

Chapter 9: Let's Give This a Pulse

1. Aldo Leopold, *A Sand County Almanac: And Sketches Here and There* (London: Oxford University Press, 1949).

— INDEX —

Index

Index

Index

Index

Index

— ABOUT THE AUTHOR —

WWW.CHRISROBBINS.CO.UK

DEREK GOW IS A FARMER AND NATURE CONSERVAtionist, and author of *Bringing Back the Beaver*. Born in Dundee in 1965, he left school when he was seventeen and worked in agriculture for five years. Inspired by the writing of Gerald Durrell, he jumped at the chance to manage a European wildlife park in central Scotland in the late 1990s before moving on to develop two nature centres in England. He now lives with his children, Maysie and Kyle, on a three-hundred-acre farm on the Devon-Cornwall border, which he is in the process of rewilding. Derek has played a significant role in the reintroduction of the Eurasian beaver, the water vole and the white stork in England. He is currently working on a reintroduction project for the wildcat and a book on the last of the British wolves.

rewildingcoombeshead.co.uk
Twitter @gow_derek
Instagram @derekgow